Bionic Sensing with Artificial Lateral Line Systems for Fish-Like Underwater Robots

In this book, the authors first introduce two fish-like underwater robots, including a multiple fins-actuated robotic fish and a caudal fin-actuated robotic fish with a barycenter regulating mechanism. They study how a robotic fish uses its onboard pressure sensor arrays-based ALLS to estimate its trajectory in multiple locomotions, including rectilinear motion, turning motion, ascending motion, and spiral motion. In addition, they also explore the ALLS-based relative position and attitude perception between two robotic fish in a leader-follower formation. Four regression methods—multiple linear regression methods, support vector regressions, back propagation neural networks, and random forest methods—are used to evaluate the relative positions or attitudes using the ALLS data.

The research on ALLS-based local sensing between two adjacent fish robots extends current research from one individual underwater robot to two robots in formation, and will attract increasing attention from scholars of robotics, underwater technology, biomechanics and systems, and control engineering.

Guangming Xie received his B.S. degrees in both Applied Mathematics and Electronic and Computer Technology, his M.E. degree in Control Theory and Control Engineering, and his Ph.D. degree in Control Theory and Control Engineering from Tsinghua University, Beijing, China in 1996, 1998, and 2001, respectively. Then he worked as a postdoctoral research fellow in the Center for Systems and Control, Department of Mechanics and Engineering Science, Peking University, Beijing, China from July 2001 to June 2003. In July 2003, he joined the Center as a lecturer. Now he is a Professor of Dynamics and Control in the College of Engineering, Peking University.

Xingwen Zheng received his Ph.D. degree in General Mechanics and Foundation of Mechanics from Peking University in 2020 and his B.E. degree in Mechanical Engineering and Automation from Northeastern University in 2015. He is currently a JSPS postdoctoral researcher at the University of Tokyo. His major research interests include robotics, biomimetics, electronics, MEMS/NEMS technology, and artificial intelligence. His current research focuses on understanding the ultrasensitive flow-sensing abilities of seal whisker and fish lateral line, then designing sensors which rival their sensing abilities, and finally applying the flow sensors on underwater robots to assist the navigation, sensing, and motion control of the robots.

Bionic Sensing with Artificial Lateral Line Systems for Fish-Like Underwater Robots

Guangming Xie
Xingwen Zheng

CRC Press
Taylor & Francis Group
Boca Raton London New York

CRC Press is an imprint of the
Taylor & Francis Group, an **informa** business

First edition published 2023
by CRC Press
6000 Broken Sound Parkway NW, Suite 300, Boca Raton, FL 33487-2742

and by CRC Press
4 Park Square, Milton Park, Abingdon, Oxon, OX14 4RN

CRC Press is an imprint of Taylor & Francis Group, LLC

© 2023 Guangming Xie and Xingwen Zheng

Library of Congress Cataloging-in-Publication Data

Names: Xie, Guangming, 1972- author. | Zheng, Xingwen, 1993- author.
Title: Bionic sensing with artificial lateral line systems for fish-like underwater robot / Guangming Xie and Xingwen Zheng.
Description: First edition. | Boca Raton : CRC Press, 2023. | Includes bibliographical references. |
Identifiers: LCCN 2022007228 (print) | LCCN 2022007229 (ebook) | ISBN 9781032316161 (hbk) | ISBN 9781032316185 (pbk) | ISBN 9781003310587 (ebk)
Subjects: LCSH: Robotic fish. | Lateral line organs--Simulation methods.
Classification: LCC TJ211.34 .X54 2023 (print) | LCC TJ211.34 (ebook) | DDC 629.8/92--dc23/eng/20220504
LC record available at https://lccn.loc.gov/2022007228
LC ebook record available at https://lccn.loc.gov/2022007229

ISBN: 978-1-032-31616-1 (hbk)
ISBN: 978-1-032-31618-5 (pbk)
ISBN: 978-1-003-31058-7 (ebk)

DOI: 10.1201/b23027

Typeset in Latin Modern font
by KnowledgeWorks Global Ltd.

Contents

CHAPTER 1

Introduction

B IO-INSPIRED FISH-LIKE ROBOT has become a focus in underwater robotics research in recent years. This book focuses on bionic sensing with artificial lateral line systems (ALLS) for fish-like underwater robot.

1.1 RESEARCH BACKGROUND

Lateral line system (LLS) is a sensory system which can be found in most species of fish. The major unit of lateral line is neuromast, which is a mechanoreceptive organ enabling fish to respond to mechanical changes in water. It consists of two kinds of neuromasts, namely, superficial neuromasts and canal neuromasts. Superficial neuromasts are situated on the surface of fish skin, while canal neuromasts are enclosed in subepidermal canals [1]. It has been demonstrated that fish can effectively detect flow velocity and pressure in the surrounding flow field using LLS [2]. Based on this characteristic, LLS serves functions in varieties of flow-aided fish behaviors, such as rheotaxis (which specifically refers to turning to face into an oncoming current), obstacle avoidance, schooling, and prey localization [3, 4].

Inspired by such excellent performances of LLS in fish behaviors, artificial lateral line research has captured more and more attention, and multiple artificial lateral line systems (ALLSs) composed of varieties of sensors have been developed in recent years [5, 6]. The existing research on artificial lateral line has mainly focused on localization of dipole source [7–10], estimation of flow characteristics [11–18] and flow-aided control of underwater robots and vehicles [14, 19–32]. Among this research, investigating the possibility of ALLS in improving performances of underwater robots and vehicles has captured increasing

DOI: 10.1201/b23027-1

1

interests. Specifically, J. Guo's group has utilized an ALLS composed of polyvinylidene fluoride (PVDF)-based pressure sensors to make a robotic fish follow an oscillating source [20]. They have also studied "wall following control" of a robotic fish using an onboard ALLS [21]. M. Kruusmaa's group has implemented fish-like rheotaxis behavior using a robotic fish with an onboard ALLS composed of pressure sensor arrays [26]. They have also conducted research on how to increase swimming efficiency of the robotic fish [27] and how to realize station holding of the robotic fish in steady flow and Kármán vortex street [28], with the aid of the ALLS. M. Triantafyllou's group has designed a MEMS pressure sensors based ALLS and mounted it on the surface of an underwater vehicle for identifying and locating underwater obstacles, thus assisting the operation and navigation of the vehicle [33]. They have also applied ALLS to benefit the situational awareness and hydrodynamic control of underwater vehicle and robot [29–32]. The above-mentioned research has demonstrated the excellent performance and great potential of ALLS in promoting technological innovation for underwater robots and vehicles. In this book, we introduce our group's studies about how a robotic fish uses its onboard pressure sensor arrays based ALLS to estimate its own motion parameters (including the velocity, trajectory, etc.) [22, 34] and sense its relative position and attitude to an adjacent robotic fish [35–37]

1.2 DOCUMENT STRUCTURE

The remainder of this article is organized as follows.

- Chapter 2 introduces the robotic fish and the onboard ALLS. Two fish-like underwater robots, including a multiple fins-actuated robotic fish and a caudal fin-actuated robotic fish with barycentre regulating mechanism, were designed. The mechanical design and electrical system of the two fish-like underwater robots were described, the test for analyzing individual differences among the pressure sensors in ALLS was conducted, the control for fins and weight block for realizing multiple three-dimensional motions was analyzed, and the recording and transmission of sensor data were investigated.

- Chapter 3 establishes the pressure variation (PV) model, which links the PVs on the surface of robotic fish and motion parameters using Bernoulli equation. The data-driven method was used for determining the model parameters using the motion parameters data and ALLS data obtained from multiple kinetics experiments of the robotic fish. Basing on the obtained PV models, the motion parameters were estimated by reversely solving the PV models using ALLS data. Then algorithms for evaluating the trajectory of fish-like robot using the estimated motion parameters were proposed, for realizing the online localization of the fish-like robot.

- Chapter 4 describes ALLS-based local sensing between two adjacent fish-like robots, for initially investigating the efficiency of ALLS in applications of a group of robotic fish. Specifically, the two robotic fish were located in a line along with the flow, with specific relative positions and attitudes in a water flume. Then multiple measurements were conducted for obtaining the regularity curves between the ALLS-measured PVs and the relative positions or attitudes. The explanations for the curves were described in detail.

- Chapter 5 investigates the regression model which links the ALLS data and relative positions or attitudes. The sensitivity of each pressure sensor to the relative positions or attitudes was analyzed, and the insufficiency and redundancy of the pressure sensors were investigated. Then four typical regression methods, including multiple linear regression method, support vector regression, back propagation neural network, and random forest method, were used for establishing the regression models. Basing on the established models, the relative positions or attitudes was obtained using the ALLS data.

- Chapter 6 concludes this book with an outline of future work.

CHAPTER 2

Fish Lateral Line Inspired Perception and Flow-Aided Control: A Review

A NY PHENOMENON IN NATURE is potential to be an inspiration for us to propose new ideas. Lateral line is a typical example which has attracted more interest in recent years. With the aid of lateral line, fish is capable of acquiring fluid information around, which is of great significance for them to survive, communicate and hunt underwater. In this chapter, we briefly introduce the morphology and mechanism of the lateral line first. Then we focus on the development of artificial lateral line, which typically consists of an array of sensors and can be installed on underwater robots. A series of sensors inspired by the lateral line with different sensing principles have been summarized. And then the applications of artificial lateral line systems in hydrodynamic environment sensing and vortices detection, dipole oscillation source detection, and autonomous control of underwater robots have been reviewed. In addition, the existing problems and future foci in this field have been further discussed in detail. The current works and future foci have demonstrated that artificial lateral line has great potentials of applications and contributes to the development of underwater robots.

DOI: 10.1201/b23027-2

2.1 INTRODUCTION

Comparing with extracting land resources, it's more difficult to exploit marine resources. Due to the complexity of underwater environment, the causticity of the seawater, the high pressure of the deep seafloor, the poor visibility in the sea, and the strong interference to the sensors, people are facing extremely harsh conditions when exploiting marine resources. With the rapid development of machine manufacturing and artificial intelligence, robots serve significant functions in resource exploitation.

However, the development of underwater robots is more difficult than that of land robots because of the above-mentioned reasons. With the further development of marine resources, the disadvantages of traditional underwater robots are gradually highlighted: large size, complex system, low control flexibility, high energy consumption, low efficiency, and lack of autonomy.

Biomimetics which acts as a new comprehensive subject provides another way for underwater robot manufacturing. Many researchers have taken inspiration from aquatic organisms and developed new underwater robots from the bionic perspective, commonly known as biomimetic underwater robots. For aquatic organisms, they have lived in marine environment for billions of years and adapted to the environment through evolution. They have advantages of high maneuverability, high sensitivity to changes in surrounding environment and high-energy efficiency, etc. Inspired by the excellent performance mentioned above, various kinds of biomimetic underwater robots have been developed successfully: robotic fish [38–41], robotic dolphin [42], robotic snake [43], robotic turtles [44], robotic salamanders [45], etc.

Owing to the complexity of underwater environment, robots must be equipped with well-developed underwater sensing system, which can help them to adapt to the environment and perform underwater tasks. Due to the particularity of water environment, land sensors cannot be directly used under the water, which restricts the development of sensing technology for underwater robots to some extent. To solve this problem, scientists have taken inspiration from marine life once again. Over billions of years of evolution in the ocean, fish have developed advanced systems for sensing the water environment. Lateral line is such a unique sensing system of fish and plays a key role for fish behaviors in complex water environment. The research and development of artificial lateral line (ALL) system referring to an array of sensors installed on the underwater vehicle will greatly improve the perception ability of the robots, which is advantageous for underwater missions.

In this chapter, we firstly describe the biology of lateral line and the models which have been established to illustrate the mechanisms in section 2.2. In section 2.3, we review the sensors that are inspired by the lateral line and based on different sensing principles according to the categories. In section 4, we present applications of ALL in hydrodynamic environment sensing and vortices detection. In section 2.5, we introduce results in dipole oscillation source detection with the help of ALL using different methods. Finally, we discuss the flow-aided control of underwater robots using ALL system for different purposes. At the end, we discuss the problems of current studies and try to put forward possible improvement strategies for further studies. The abbreviations used and corresponding full names are listed in Table 2.1.

TABLE 2.1 The Abbreviations Used and Corresponding Full Names

Abbreviation	Full name
AlN	Aluminium Nitride
AHC	Artificial hair cell
ALL	Artificial lateral line
CN	Canal neuromast
CRB	Cramer-Rao bound
CRLB	Cramer-Rao lower bound
HWA	Hot-wire anemometer
IMU	Inertial measurement unit
IPMC	Ionic polymer-metal composite
LCP	Liquid crystal polymer
MEMS	Micro-electromechanical systems
MMSE	Minimum mean-squared error
MVDR	Minimum variance distortionless response
MUSIC	Multiple signal classification
PDMA	Plastic deformation magnetic assembly
PDMS	Polydimethylsiloxane
PVC	Polyvinyl chloride
PVDF	Polyvinylidene fluoride
SOI	Silicon-On-Insulator
SN	Superficial neuromast
SMD	Surface-mounted devices

2.2 MECHANISMS AND MODELS OF THE FISH LATERAL LINE

Lateral line is the most highly differentiated structure in the sensory organs of the skin, which is typically sulcus or tubular. It is a sensory organ peculiar to fish and aquatic amphibians and important for them to maneuver in the darkness [46]. The sensory unit of the lateral line is the neuromast, a receptor that consists of sensory hair cells and support cells [47]. There are two types of lateral line neuromasts: superficial neuromast (SN) and canal neuromast (CN), as shown in Figure 2.1. Both of these two kinds of neuromasts sense the stimulation generated by water flow through sensory cells. Due to the difference in distribution, number and morphology of sensory cells, the two kinds of neuromasts have different functions [48].

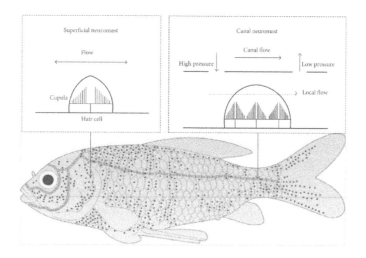

Figure 2.1 Lateral line and neuromasts of a fish. Black dots represent locations of SNs and white dots show the approximate locations of canal pores [49].

The SNs which act as displacement sensors are free-standing on the skin or on pedestals grown above the skin. They are usually located in lines on the fish body [50]. The ability of sensing the flow direction and velocity is mainly realized by SNs which are sensitive to the displacement and respond to the low-frequency direct current component. When the water flow and the fish surface move relatively to each other, the SNs bend, causing the neuromasts below to produce nerve impulses which will be transmitted from nerve endings to the nerve centers of the brain. Under such a mechanism, fish can sense the flow information with the help of SNs. On the other hand, CNs are equivalent to pressure gradient sensors which can sense pressure gradient and are sensitive to acceleration and respond to high-frequency components. The CNs are located in the lateral line canals that are full of mucus under the epidermis of fish and communicate with the external water environment through some small holes [51]. When there is a velocity gradient between adjacent holes, the pressure difference will be generated, leading to the fluid movement in the lateral line canals, which triggers the nerve impulse. The lateral line system composed of CNs and SNs can sense various stimuli from different directions, so that fish can acquire enough information for a full sense of surrounding water environment.

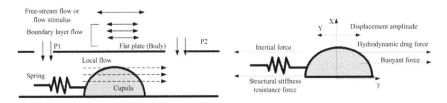

Figure 2.2 Biomechanical model and force analysis for a CN [52].

Scientists have established theoretical models to describe the sensing mechanisms of these neuromasts and to study the interactions between flow and fish, which is instructive to the development of ALL system. As shown in Figure 2.2, the CN can be regarded as a rigid hemisphere sliding over a frictionless plate in biomechanics which is coupled with a linear spring [52,53]. For the forces on the CNs, the governing equation can be described as:

$$F_m + F_k = F_u + F_b \tag{2.1}$$

The terms on the left are the inertial force and the structural stiffness resistance which can be written as the following forms respectively:

$$F_m = M\frac{d^2Y(t)}{dt^2} \tag{2.2}$$

$$F_k = KY(t) \tag{2.3}$$

where M represents the cupula mass, $Y(t)$ represents the displacement induced by the passing flow, K is the sliding stiffness. The terms on the right are the hydrodynamic drag force and buoyant force due to the pressure difference which can be defined as the following forms:

$$F_u = D\left[\frac{dY(t)}{dt} - \frac{dW(t)}{dt}\right] \tag{2.4}$$

$$F_b = M\frac{d^2W(t)}{dt^2} \tag{2.5}$$

where D is the drag coefficient, $W(t)$ is the displacement of external flow. Thus $\frac{dY(t)}{dt} - \frac{dW(t)}{dt}$ represents the relative velocity of the cupula and external flow. According to the above results, the governing equation can be written as

$$M\frac{d^2Y(t)}{dt^2} + KY(t) = D\left[\frac{dY(t)}{dt} - \frac{dW(t)}{dt}\right] + M\frac{d^2W(t)}{dt^2} \tag{2.6}$$

In order to describe the sensitivity of CNs, considering that the equation is linear, the displacement can be decomposed into different frequencies. At a certain frequency, $Y(t)$ and $W(t)$ are the steady-state oscillatory displacement and the oscillating flow displacement with a form of $Y(t) = Y_0(f)e^{i2\pi ft}$ and $W(t) = W_0(f)e^{i2\pi ft}$. So the velocity of the flow is expressed as

$$V(t) = \frac{dW(t)}{dt} = i2\pi fW_0(f)e^{i2\pi ft} = V_0(f)e^{i2\pi ft} \tag{2.7}$$

The frequency-dependent sensitivity of the CNs is defined as the ratio of displace amplitude of cupula and the velocity amplitude of flow, which is shown in the following formula:

$$S_{CN}(f) = \frac{Y_0(f)}{V_0(f)} = \frac{1}{-2\pi f_t}\frac{1 + \frac{\sqrt{2}}{2}(1+i)\sqrt{\frac{f}{f_t}} + \frac{1}{3}i\frac{f}{f_t}}{N_r + i\frac{f}{f_t} - \frac{\sqrt{2}}{2}(1-i)(\frac{f}{f_t})^{\frac{3}{2}} - \frac{1}{2}(\frac{f}{f_t})^2} \tag{2.8}$$

where f_t is the transition frequency expressed as $f_t = \frac{\mu}{2\pi\rho_w a^2}$, which determine viscous ($f < f_t$) or inertial ($f > f_t$) forces dominate the fluid forces applied on the cupula. μ is the dynamic viscosity of fluid. $N_r = \frac{Ka\rho_w}{6\pi\mu^2}$, which represents the resonance properties, is the resonance factor.

Different from the CN, the SN is modeled as two connecting beams with different bending rigidity, as shown in Figure 2.3. The distal beam is more flexible than the proximal beam because of the difference in material properties of the proximal and distal parts of cupula. A spring is used to simulate the torsional stiffness generated by the hair beam.

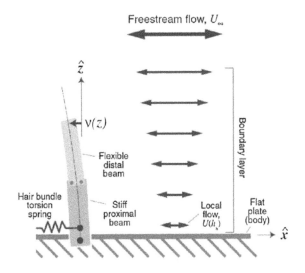

Figure 2.3 Biomechanical model of a SN [54].

The governing equation is different from that of CNs, forces applied on the beams are also functions of elevation z due to the uneven distribution of forces when a beam bends. The equation is expressed as

$$F_m(z) + F_e(z) = F_u(z) + F_a(z) + F_b(z) \qquad (2.9)$$

The terms from left to right represent the inertial force, elastic stiffness term, hydrodynamic drag force, acceleration reaction force, and buoyant force respectively. Similar to CNs, the sensitivity of SNs is defined as the ratio of cupula deflection $v(H)$ at the height of the beam and free-stream velocity U_∞, which is written as

$$S_{SN}(f) = \frac{v(H)}{U_\infty} = -\frac{ib_w}{2\pi f b_m}\left[1 - \frac{i\pi f b_m \delta^4}{2EI + i\pi f b_m \delta^4}e^{-\frac{H(1+i)}{\delta}}\right]$$

$$+ \sum_{j=0}^{3} C_j e^{ij H^4}\sqrt{\frac{2\pi f i b_m}{EI}} \qquad (2.10)$$

where C_j represents a sequence of four integration constants. H is the height of the top of the beam. EI is the bending modulus of the beam. δ is the thickness of boundary layer expressed as $\delta = \sqrt{\frac{2\mu}{\rho_w \omega}}$, where ω is the angular speed of the stimulus. More details on the parameters b_m and b_w can be referred to the paper [54].

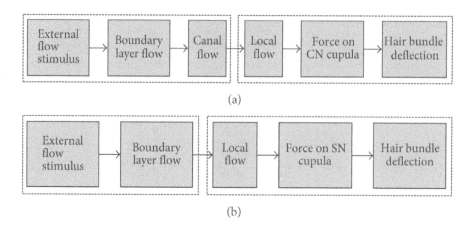

(a)

(b)

Figure 2.4 Propagation path of lateral line neuromasts. (a) Propagation path of CNs. (b) Propagation path of SNs [52].

Figure 2.4 shows the propagation path of lateral line neuromasts. On one hand, for CNs, the propagation path is divided into two steps. At the first step, the velocity or acceleration of the external free flow is converted into the velocity of the local flow with the help of boundary layer and canals. Flow outside induces the pressure difference between the canals, which triggers flow velocity inside the canal. At the second step, canal flow applies fluid forces on the cupula and results in the CN deflection. On the other hand, for SNs, at the first step, the velocity or acceleration of the external free flow is converted into the velocity of the local flow similarly without the reflection of canal flow. At the second step, SNs deflects under the forces induced by local flow [52].

With the assistance of SNs and CNs, fish can detect the flow direction, speed, and pressure gradients respectively. Moreover, SNs can distinguish fields in a spatial uniform flow and in a turbulent flow, while CNs only respond to a non-uniform flow field, such as the fluctuation of water produced by a vibrating sphere or a swimming fish.

2.3 THE EXISTING ALL SENSORS AND SYSTEMS

2.3.1 ALL Sensor Unit

As mentioned above, scientists have established mathematical models to interpret mechanisms that how fish acquire fluid information assisted by lateral line. The results can be an inspiration of ALL. Owing to limitations of existing technologies for underwater detection such as scattering and multipath propagation issues for acoustic sensors and turbidity of the sea for optical sensors [52], varieties of fish sensing organs inspired sensors were developed using different principles. Considering that commercial pressure sensors are not as sensitive as the lateral line receptors and functions are limited, various kinds of self-developed ALL sensors have been explored. The existing ALL sensors are based on different sensing mechanisms, including piezoresistive, piezoelectric, capacitive, optical, thermal, and magnetic effect. The research status of ALL sensors with different sensing mechanisms will be discussed below.

2.3.1.1 Piezoresistive ALL Sensors

Piezoresistive sensor is a device based on the piezoresistive effect of the semiconductor material on the substrate. Piezoresistive effect refers to a phenomenon that the electrical resistance of the material changes while it is subjected to force. As a result, the bridge on the substrate produces corresponding unbalanced output. In this way, the substrate can be directly used as an element to measure pressure, tension, etc. Based on the quantity measured directly, the information about the environment is available. Figure 2.5 shows various piezoresistive ALL sensors mentioned below.

In 2002, Fan *et al.* first made a major breakthrough in piezoresistive ALL sensors fabrication using combined bulk micromachining methods and an efficient three-dimensional assembly method, namely, plastic deformation magnetic assembly (PDMA) process. They leveraged PDMA to realize the vertical cilium, which was important in the hair cell. A single sensor was composed of an in-plane fixed-free cantilever mainly made of Boron ion diffused Si using etching technology, a vertical artificial cilium attached at the free end and a strain gauge located at the base of the horizontal cantilever which was used to sense the bending of the vertical cilium. Subjected to the impact of local flow, the vertical cilium bent and transferred the influence to the cantilever beam. The corresponding results were measured by the strain gauge. The sensors were used to detect laminar flows ranging from 0.1 to 1 m/s [55].

In 2003, Chen *et al.* compared the above sensor with the hot-wire anemometer and improved the design by rigidly connecting the vertical hair to the substrate and placing the strain gauge at the root of the hair directly, which made an advance in the spatial resolution of the sensor [65]. In order to further improve the sensitivity and resolution,

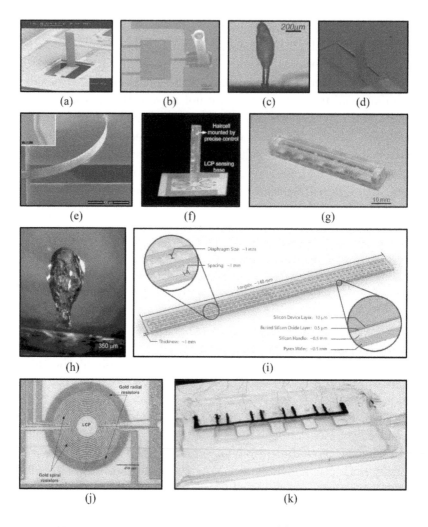

Figure 2.5 Various piezoresistive ALL sensors. (a) Scanning electron micrograph of a single artificial hair cell sensor [55]. (b) Scanning electron micrograph of an AHC sensor [56]. (c) The front-view of a hair sensor after being coated with the hydrogel material [57]. (d) Scanning electron micrograph on a bent cantilever [58]. (e) Scanning electron micrograph on a fabricated cantilever [59]. (f) An angle view of the complete sensor with hair cell [60]. (g) Photographic image of the final ALL canal system prototype [16]. (h) Cupula formed using nanofibrils scaffold has a prolate spheroid shape [61]. (i) Diagram of the pressure sensor array with basic structure depicted [62]. (j) Photograph of the pressure sensor array [63]. (k) Optical microscopic image of a fabricated full-bridge LCP sensor with two radial and two spiral gold piezoresistors [64].

in 2007, Yang *et al.* proposed another highly sensitive piezoresistive flow sensor fabricated on a silicon-on-insulator (SOI) wafer, the cilium of which was made of photodefinable SU-8 epoxy and adopted a symmetric cylindrical shape [66]. The sensor was used to detect the steady-state laminar flow and oscillatory flow with a threshold down to 0.7 mm/s [56, 66, 67]. In 2010, they assembled piezoresistive sensors on the surface of a cylindrical polyvinyl chloride (PVC) model and put forward an adaptive beamforming algorithm in order to locate the dipole source [7].

To match the threshold sensitivity of the integrated fish flow sensory system, McConney *et al.* created a bio-inspired hydrogel-capped hair sensory system in 2008 using a precision drop-casting method. They added extremely compliant and high-aspect-ratio hydrogel cupula (polyethylene glycol) to the SU-8 hair sensors, as a result of which, the sensitivity of the sensors was enhanced by about two orders of magnitude (2.5 μm/s) [57].

Unlike the sensors mentioned above whose material was mainly silicon, Qualtieri *et al.* reported on a kind of ALL sensor in 2011, the key component of which was the stress-driven aluminium nitride (AlN) cantilevers. The structures utilizing a multilayered cantilever AlN/Molybdenum (Mo) and a Nichrome 80/20 alloy as piezoresistor were realized by means of micromachining techniques combining optical lithography and etching process. The piezoresistor showed a sensitivity to directionality and low value pressure with a detection threshold of 0.025 bar [58]. Besides, in 2012, they deposited a water resistant parylene conformal coating on the hair cell and developed a biomimetic waterproof Si/SiN multilayered cantilever using surface micromachining techniques. The sensor showed mechanical robustness in high-speed flow and had the capability of discriminating the flow direction at low frequencies [59].

In 2014, Kottapalli *et al.* developed an artificial SN sensor array composed of a liquid crystal polymer (LCP) membrane, a gold strain gauge and a Si-60 cilium fabricated by stereolithography with a high-aspect ratio of 6.5. The sensors demonstrated a high sensitivity of 0.9 mV/(m/s) and 0.022 V/(m/s) while detecting air and water flows. And the threshold velocity limits were 0.1 m/s and 15 mm/s, respectively [60]. In 2016, they created a canopy-like nanofiber pyramid around the Si-60 polymer cilium and then dropped the casting hydrogel cupula onto the nanofiber scaffold to enhance the sensors. The velocity detection threshold of the sensor was 18 mm/s while it is used in water flow sensing [61].

All sensors mentioned above feature a cilium and a cantilever beam, which bends under the impact of water flow and is sensitive to flow velocity. Except for this, other piezoresistive sensors are mostly planar and the piezoresistors are directly installed on the substrate to detect underwater pressure distribution and variations.

In 2017, Jiang *et al.* integrated cantilevered flow-sensing elements mainly made of polypropylene and polyvinylidene fluoride (PVDF) layers in a polydimethylsiloxane (PDMS) canal and used it to detect a dipole vibration source. The sensors showed high-pass filtering capability and a pressure gradient detection limit of 11 Pa/m at the frequency of 115 Hz [16].

Additionally, Vicente *et al.* used an array of off-the-shelf pressure sensors to detect cylindrical obstacles of round and square in 2007. The array consisted of hundreds of micro-electromechanical systems (MEMS) pressure sensors which were fabricated on etched Silicon and Pyrex wafers. A strain gauge was mounted on a flexible diaphragm which was a thin (20 μm) layer of Silicon attached at the edges of a square silicon cavity with a width of 2000 μm and served as the sensing element with a pressure detection threshold of 1 Pa [62]. In 2012, they presented a 1-D array of four sensors with a 15-mm center-to-center spacing. Each sensor had two key components: a strain-concentrating PDMS diaphragm and a resistive strain gauge made of a conductive carbon-black PDMS composite. And the resolution of it was 1.5 Pa [63].

To perform underwater surveillance, Kottapalli *et al.* developed an array of polymer MEMS pressure sensors fabricated with a Cr (20 nm)/Au (700 nm) thick gold layer sputtered on a flexible substrate and LCP serving as the sensing membrane material. Installed on curved surfaces of the underwater vehicle bodies, the sensor detected underwater objects by sensing the pressure variations. Compared with silicon-based hair vertical structures or thin metal cantilever beams, it showed a better sensitivity of 14.3 μV/Pa and a better resolution of 25 mm/s in water flow sensing [64].

2.3.1.2 *Piezoelectric ALL Sensors*

Piezoelectricity refers to the electric charge generated on the surface of some certain materials while subjected to external forces. This effect inspires another kind of ALL sensors which is able to sense environment by collecting the electric information. Figure 2.6 shows various piezoelectric ALL sensors mentioned below.

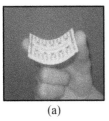
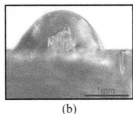
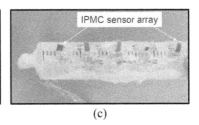

(a) (b) (c)

Figure 2.6 Various piezoelectric ALL sensors. (a) Array of 2 by 5 piezoelectric sensors on flexible LCP substrate [33]. (b) Microscopic side-view image of the sensor showing the hydrogel cupula and the PDMS pillars with height gradient [68]. (c) The IPMC-based lateral line prototype [69].

For the performance of situational awareness and obstacle avoidance, Asadnia *et al.* used floating bottom electrode to design an array of $Pb(Zr_{0.52}Ti_{0.48})O_3$ thin-film piezoelectric pressure sensors in 2013 [33]. Packaged into an array of 25 sensors on a flexible liquid crystal polymer substrate patterned with gold interconnects, the array was used to locate a vibrating sphere dipole in water and showed a resolution of 3 mm/s in detecting oscillatory flow velocity. Besides, the sensors had many advantages such as self-powered, miniaturized, light-weight, low-cost, and robust. In 2015, they optimized the sensor by mounting a stereolithographically fabricated polymer hair cell on microdiaphragm with floating bottom electrode. The sensors demonstrated a high-pass filtering nature with a cut-off frequency of 10 Hz, a high sensitivity of 22 mV/(mm/s), and a resolution of 8.2 mm/s in water flow detection [70]. In 2016, they reported the development of a new class of miniature all-polymer flow sensors with an artificial ciliary bundle fabricated by combining bundled PDMS micro-pillars with graded heights and electrospinning PVDF piezoelectric nanofiber tip links. By means of precision drop-casting and swelling processes, a dome-shaped hyaluronic acid hydrogel cupula encapsulating the artificial hair cell bundle was formed. The sensors achieved a sensitivity of 300 mV/(m/s) and a threshold detection limit of 8 μm/s respectively [68].

In 2011, Abdulsadda *et al.* proposed a novel ALL sensors utilizing the inherent sensing capability of ionic polymer-metal composites (IPMCs). An IPMC consisted of three layers, with an ion-exchange polymer membrane sandwiched by metal electrodes. Detectable electrical signals were produced under the impact of external forces. The IPMC flow sensor were used to localize dipole sources 4-5 body away and demonstrated a threshold detection limit of less than 1 mm/s [69].

2.3.1.3 Capacitive ALL Sensors

Owing to the high sensitivity and low-power consumption, capacitance principle has been widely used in many different types of sensors. The key component is the capacitive readout which has the capability of converting external stimulus into capacitance changes, which provides an effective way to detect underwater pressure and flow velocity. Like the piezoresistive sensors, the hair attached to the membrane will respond to the impact from the local flow. And then the membrane reflects and changes the distance or gap between the electrodes. As a result, the change in capacitance is quantitatively related to the external impact. Figure 2.7 shows various capacitive ALL sensors mentioned below.

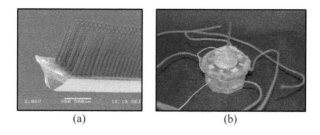

(a)   (b)

Figure 2.7   Various capacitive ALL sensors. (a) Scanning Electron Microscope image of actual sensors [71]. (b) Top views of the ALL sensor [72].

In 2007, Krijnen *et al.* reported developments in hair sensors based on mechanoreceptive sensory hairs of crickets using artificial polysilicon technique to form Silicon Nitride-suspended membranes and SU-8 polymer processing for hairs with diameters of about 50 μm and up to 1 mm in length. The membranes have thin chromium electrodes on top which formed variable capacitors and the sensitivity of the sensor is 1.39 pF/rad [71, 73]. In 2010, they realized the dense arrays of fully supported flexible SU-8 membranes with integrated electrodes underneath that supported cylindrical hair-like structures on the top. While used in air flow detection, the mechanical sensitivity at the frequency of 115 Hz was 0.004 rad/(mèûŕs) [74].

Another capacitive whisker sensor inspired by seal vibrissae was developed by Stocking *et al.* to measure the flow velocity and detect the direction in 2010. They mounted a rigid artificial whisker on a novel cone-in-cone parallel-plate capacitor base which was covered by a PDMS membrane. Numerical simulation predicted the change of capacitor output signal in a range of 1 pF when the flow velocity varied from 0 to 1.0 m/s [72].

2.3.1.4 *Optical ALL Sensors*

Optical principles have also been used to develop ALL sensors. Figure 2.8 shows various optical ALL sensors mentioned below. Klein *et al.* made a great breakthrough in this area in 2011. The artificial canal neuromasts segment they developed consisted of a transparent silicone bar which had the same density of water and an infrared light emitting diode at one end of the silicone bar. To detect the fluid motion, light, leaving the opposite end of the silicone bar, illuminated an optical fiber that was connected to a surface-mounted devices (SMD) phototransistor and the output was amplified and converted to be stored on a computer. The sensors are used to detect water movements caused by a stationary vibrating sphere or a passing object and vortices caused by an upstream cylinder. According to the acquired information, they calculated the bulk flow velocity and the size of the cylinder producing the vortices. The detection limit in water flow was 100 μm/s [75].

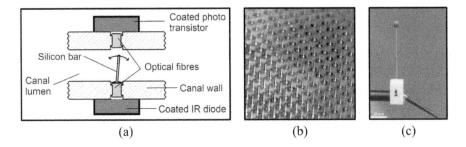

Figure 2.8 Various optical ALL sensors. (a) Scheme of an artificial CN [75]. (b) Scanning-Electron Microscope image of a pillar array [76]. (c) The photograph of the all-optical sensor [77].

Another method was put forward by Wolfgang *et al.* in 2009. The key component of the sensor was flexible micro-pillars which protruded into local flows and bent subjected to exerted drag forces. The pillar was fabricated from the elastomer PDMS and the deflection was measured by means of optical methods [76].

In 2018, Wolf *et al.* presented an all-optical 2D flow velocity sensor consisting of optical fibres inscribed with Bragg gratings supporting a fluid force recipient sphere. The artificial neuromast demonstrated a threshold of 5 mm/s at a low frequency and 5 μm/s at resonance with a typical linear dynamic range of 38 dB at 100 Hz sampling. Additionally, the artificial neuromast is capable of detecting flow direction within a few degrees [77].

2.3.1.5 Hot-Wire ALL Sensors

Hot-wire anemometer (HWA) uses a heated wire placed in the air. While air or water flows through it, heat loss leads to changes in temperature and resistance. Therefore, we can measure the velocity by detecting electrical signals. Figure 2.9 shows various hot-wire ALL sensors mentioned below.

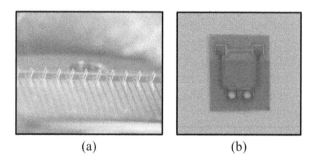

(a)   (b)

Figure 2.9 Various hot-wire ALL sensors. (a) An optical micrograph of an ALL [78]. (b) A sensor unit [79].

In 2006, Yang *et al.* developed a surface-micromachined, out-of-plane ALL sensor array using the principle of thermal HWA. Inspired by the SNs of fish, the hot wire was lifted from the substrate by two pointed heads. They utilized photolithography technique to fabricate the sensor in plane and assembled it outside by three-dimensional magnetic assembly. The sensor afterwards was used to track the position of a vibrating dipole source and exhibited a threshold of 0.2 mm/s with a bandwidth of 1 kHz [78,80,81].

Liu *et al.* proposed a novel micromachined hot-film flow sensor system realized by using a film depositing process and a standard printed circuit in 2009. They preprinted the sensor electrodes and electronic circuits on a flexible substrate of polyimide and utilized Cr/Ni/Pt as the sensing element with a resistance temperature coefficient around 2000 ppm/K. The resolution of this sensor was 0.1 m/s [79].

2.3.2 ALL Sensors Placement Optimization

With the development of ALL sensors unit, more efforts have been devoted to the sensors placement optimization in order to simulate the fish lateral line to a greater extent. Verma *et al.* used a larva-shaped swimmer exposed in disturbances induced by oscillating, rotating and cylinders to conduct experiments in 2020. Combining Navier-Stokes equations with Bayesian experimental design and with a purpose of detecting the location of the source, they presented that shear sensors should be installed on the head and the tail while pressure sensors should be distributed uniformly along the body and intensively on the head, which is similar to real fish lateral line [82]. In 2019, Xu *et al.* put forward an optimal weight analysis algorithm combined with feature distance and variance evaluation and 3 indexes to evaluate the performance of the sensor array. They also briefly discussed the optimal number of sensors [83]. This work has provided new ideas for studies in ALL sensors distribution optimization in the future.

In this section, we have presented ALL sensors based on different sensing mechanisms and briefly introduced works in sensors placement optimization. Table 2.2 summarizes the parameters of the ALL sensors, including transduction mechanism, processing technique, material, and sensitivity. Though great progress has been made in the design and fabrication of ALL sensors, the sensors mentioned above are mostly simple imitations of the real fish lateral line and have a long way to go in sensitivity. Firstly, the sensitivity of a single sensory unit can be further improved. Additionally, the neuromasts distribution of fish is continuously optimized in evolution. Thus, it is necessary for us to find out the optimal distribution of ALL sensors according to different shapes of robotic fish in order to simulate the lateral line to the maximum extent possible and develop a complete ALL system based on existing local sensor arrays. Last but not least, we need more efficient signal processing methods to take full advantage of the information acquired by the ALL system and make decisions like a living body. ALL system like this has great potential for future oceanographic research.

TABLE 2.2 Summary of ALL Sensors

Transduction mechanism	Author	Processing technique	Material	Sensitivity
Piezoresistive	Fan et al. 2002 [55] Chen et al. 2003 [65]	PDMA for vertical cilium, micromachining	Boron ion diffused Si for piezoresistor, Metal-permalloy for hair	100 mm/s
	Chen et al. 2007 [56] Yang et al. 2007 [66] Chen et al. 2006 [67] Yang et al. 2010 [7]	Ion implantation, deep reactive ion etching	Boron ion diffused Si for piezoresistor, SU-8 epoxy for hair	0.1 mm/s
	McConney et al. 2008 [57]	Photo polymerization	Boron ion diffused Si for piezoresistor, SU-8-hydrogel for hair	2.5 μm/s
	Qualtieri et al. 2011 [58]	Micromachining	Aluminum Ni for piezoresistor, Nichrome alloy for hair	0.025 bar
	Qualtieri et al. 2012 [59]	Micromachining	Si/SiN for piezoresistor, Parylene for hair	50 mm/s
	Kottapalli et al. 2014 [60] Kottapalli et al. 2016 [61]	Deep reactive ion etching, electrostatic spinning	LCP for membrane, gold for strain gauge, Si-60 for hair, HA-MA hydrogel for cupula	100 mm/s (air), 18 mm/s (water flows)
	Jiang et al. 2017 [16]	Micromachining	PDMS for canal, Polypropylene and PVDF for piezoresistor	11 Pa/m

	Vicente et al. 2007 [62]	Micromachining	Si	1 Pa/m
	Vicente et al. 2012 [63]	Micromachining	PDMS for diaphragm, a conductive carbon-black PDMS composite for strain gauge	1.5 Pa
	Kottapalli et al. 2012 [64]	Micromachining	LCP for membrane, gold for piezoresistors	25 mm/s
Piezoelectric	Asadnia et al. 2013 [33] Asadnia et al. 2015 [70]	Micromachining and the sol-gel method	$Pb(Zr_{0.52}Ti_{0.48})O_3$ for membrane, Si-60 for hair	3 mm/s
	Asadnia et al. 2016 [68]	Precision drop-casting and swelling processes	PDMS for micro-pillars, PVDF for tip links, HA-MA hydrogel for cupula	8 μm/s
	Abdulsadda et al. 2011 [69]	Micromachining	IPMC	Localization accuracy at the source-sensor separation of 1 body length

(Continued on next page)

TABLE 2.2 (Continued)

Transduction mechanism	Author	Processing technique	Material	Sensitivity
Capacitive	Krijnen et al. 2007 [71] Krijnen et al. 2010 [74] Krijnen et al. 2003 [73]	Sacrificial poly-silicon technology, SU-8 polymer processing	Silicon-nitride for membranes, SU-8 polymer for hair	0.004 rad/(m·s)
	Stocking et al. 2010 [72]	Micromachining	PDMS for membrane	N/A
Optical	Klein et al. 2011 [75]	N/A	Silicone for transparent bar, an infrared light emitting diode	100 μm/s
	Sebastian et al. 2009 [76]	N/A	PDMS for the pillar	N/A
	Wolf et al. 2018 [77]	N/A	N/A	5 μm/s (resonance) and 5 mm/s (low frequency)
Hot-Wire	Yang et al. 2006 [80] Pandya et al. 2006 [78] Chen et al. 2006 [81]	PDMA for vertical cilium, micromachining	Pt/Ni/Pt film for the thermal element, polyimide for support beams	0.2 mm/s
	Liu et al. 2009 [79]	Film depositing process and standard printed circuit	Polyimide for substrate, Cr/Ni/Pt for sensing element	0.1 m/s

2.4   HYDRODYNAMIC ENVIRONMENT SENSING AND VORTICES DETECTION

Due to the limitations of existing measurements methods in complex natural flows and the emergence of the above mentioned various types of sensors, scientists have started to use ALL systems consisting of these sensors to obtain information from fluid environment. Efforts are devoted into the following fields: flow field characteristics identification, flow velocity and direction detection, vortex street properties detection. Carriers boarded with ALL mentioned in this section are shown in Figure 2.10.

2.4.1   Flow Field Characteristics Identification

In 2013, Kruusmaa *et al.* presented that the robotic fish boarded with 5 pressure sensors (Figure 2.10(b)) was able to identify the flow regimens (uniform flow and periodic turbulence) according to the pressure around the underwater vehicle in 2013 [85].

In 2018, they used the ALL probe consisting of 11 piezoresistive pressure sensors (Figure 2.10(d)) to capture hydrodynamic information. The probe provided an average flow velocity in turbulent flows which was comparable to the results measured practically. According to the characteristics of the fluid, the flows were classified by comparing the probability distribution of turbulent pressure fluctuation [87].

Liu *et al.* also made great contribution in flow field characteristics identification. In 2019, they proposed an improved pressure distribution model to calculated the pressure around the ALL which consisting of 23 pressure sensors (Figure 2.10(e)) and then established a visualized pressure difference matrix to identify flow field in different conditions. A four-layer convolutional neural network model was constructed to evaluate the accuracy of this method [88].

2.4.2   Flow Velocity and Direction Detection

In 2013, Kruusmaa *et al.* proved that the flow speed can be estimated only by the average pressure on the sides of the robotic fish (Figure 2.10(b)). They put forward a fitted formula representing the relationship between the average pressure drop and the flow speed based on Bernoulli formula. Besides, they also reported that the flow direction could be detected because that the pressure on the side which the probe turned towards the flow was higher [85].

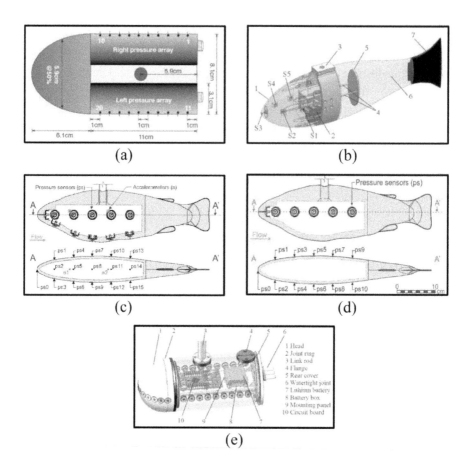

Figure 2.10 Different carriers boarded with ALL mentioned in Section 2.4. (a) A schematic diagram of the sensor platform used in [84]. (b) CAD view of the robot used in [85]: 1, rigid head of the robot; 2, servomotor; 3, middle part for holding the head and the tail; 4, steel cables; 5, actuation plate; 6, compliant tail; 7, rigid fin; S1-S5, pressure sensors. (c) Location of the 16 pressure sensors and 2 accelerometers in the ALL Probe used in [86]. (d) Illustration of the lateral line probe used for field measurements showing shape and sensor distribution in [87]. (e) A 3-D model of the carrier in [88].

In the same group, they presented a new way for flow speed estimation with the help of ALL probe consisting of 10 piezoresistive pressure sensors (Figure 2.10(c)) without sensor calibration in 2015, which is of great convenience. Induced by the interactions between fluid and the robot body, fluctuations in the pressure field around the body could be detected by the ALL probe. Based on Bernoulli formula likewise, they introduced a semiempirical resampling process. Compared with results measured by an acoustic Doppler velocimeter in a vertical slot fishway, the accuracy of this method was validated [86].

Besides, Tuhtan *et al.* made great progress in natural flow measurements using an ALL probe combined with signal processing methods in 2016. The probe is the same as showed in Figure 2.10(c). They proved that information acquired by the probe was transformed into two important hydrodynamic primitives, bulk flow velocity and bulk flow angle via canonical signal transformation and kernel ridge regression. Moreover, they showed that this method was effective not only when the sensor was parallel to the flow, but also in the condition that the angular deviation was large. While used in a natural river environment, the method had an error of 14 cm/s [89]. In 2016, they presented a new method to estimate the flow velocity ranging from 0 to 1.5 m/s. They collected time-averaged flow velocity and pressure acquired by the ALL in highly turbulent flow and put forward a signal processing approach combining Pearson product-moment correlation coefficient features and artificial neural network [90]. This method is potential to interpret the underwater preferences of fish in real environment.

Additionally, Liu *et al.* in 2020, based on a fitting method and a back propagation neural network model, successfully predicted the flow velocity and direction and the moving velocity [91]. This progress provided another possibility to identify hydrodynamic information for further study.

2.4.3 Vortex Street Properties Detection

In 2006, Yang *et al.* used ALL system to study the spatial velocity distribution of Kármán vortex street and visualized the velocity distribution of Kármán vortex street generated by a cylinder for the first time [80].

Ren *et al.* theoretically studied the perception of vortex streets using real lateral line in 2010. Based on potential flow theory, they constructed the model of flow field around the fish, and then explained how the fish captured the characteristics of vortices with the help of lateral line CNs.

The model were applied to estimate the range of the vortex, transmission speed, direction, distance between the vortex streets and distance between the fish and vortex street [92].

Klein *et al.* used an artificial canal equipped with optical flow sensors which have been presented in section 2.2 (Figure 2.8(a)). They have demonstrated the capability of the ALL canals to detect the vibrating sphere. Additionally, vortices generated by an upstream cylinder were also detected. Based on the hydrodynamic information acquired by the ALL canal, they succeeded in calculating the flow velocity and the size of the cylinder [75].

In 2012, Venturelli *et al.* used digital particle image velocimeter to visualize the flow state and a rigid body equipped with 20 pressure sensors (Figure 2.10(a)) parallel distributed to acquire flow field information and then applied time and frequency domain methods to describe hydrodynamic scenarios in steady and unsteady flows respectively. The array of pressure sensors showed a capability of discriminating vortex streets from steady flows and detecting the position and direction of the body relative to the incoming flow. A series of hydrodynamic parameters were also calculated, such as vortex shedding frequency, vortex travelling speed, and downstream distance between vortices [84].

Free *et al.* presented a method to estimate the parameters of vortices in 2017. They used a straight array of 4 pressure sensors for a spiral vortex and a square array for a Kármán vortex street. Based on potential flow theory and Bernoulli principle, the measurement equation was incorporated in a recursive Bayesian filter, as a consequence of which, the position and strength of vortices have been successfully estimated. Moreover, they identified an optimal path for underwater vehicles to swim through a Kármán vortex street using empirical observability. Experiments demonstrated the effectiveness of the closed-loop control [93]. Based on the results above, in 2018, they installed the array on a Joukowski foil and detected Kármán vortex streets nearby. With the help of trajectory-tracking feedback control, the robotic foil performed fish-like slaloming behavior in a Kármán vortex street [94].

In this section, we have focused on the application of ALL system in detecting flow characteristics. Table 2.3 as follows lists different projects mentioned above and related ongoing studies. Existing results are mainly based on static ALL sensors and experiments are conducted in laboratory environment. The characteristics of flow which can be detected are also limited. For further study, with the improvement of ALL system, we can pay more attention to natural environment experiments, where the

TABLE 2.3 Classification of Existing Studies in Hydrodynamic Environment Sensing and Vortices Detection

Project	Author	ALL sensors	Laboratory experiment/ natural environment experiment
Flow field characteristics identification	Kruusmaa et al. 2013 [85]	5 pressure sensors (Intersema MS5407-AM)	Laboratory experiment
	Kruusmaa et al. 2018 [87]	11 pressure sensors (SM5420C-030-A-P-S)	Laboratory experiment
	Liu et al. 2019 [88]	23 pressure sensors (MS5803-07BA)	Laboratory experiment
Flow velocity and direction detection	Kruusmaa et al. 2013 [85]	5 pressure sensors (Intersema MS5407-AM)	Laboratory experiment
	Kruusmaa et al. 2015 [86] Kruusmaa et al. 2016 [90]	16 pressure sensors (SM5420C-030-A-P-S) and 2 three-axis accelerometers (ADXL325BCPZ)	Laboratory experiment
	Kruusmaa et al. 2016 [89]		Laboratory experiment and natural environment experiment

(Continued on next page)

TABLE 2.3 (Continued)

Project	Author	ALL sensors	Laboratory experiment/ natural environment experiment
	Liu et al. 2020 [91]	23 pressure sensors (MS5803-07BA)	Laboratory experiment
Vortex street properties detection	Yang et al. 2006 [80]	16 HWA sensors	Laboratory experiment
	Ren et al. 2010 [92]	Theoretical model	
	Klein et al. 2011 [75]	Optical sensors	Laboratory experiment
	Venturelli et al. 2012 [84]	20 pressure sensors	Laboratory experiment
	Kruusmaa et al. 2013 [85]	5 pressure sensors (Intersema MS5407-AM)	Laboratory experiment
	Free et al. 2017 [93] Free et al. 2018 [94]	An array of 4 pressure sensors	Laboratory experiment

complexity of the water environment and the complex movements of the robot fish make it more difficult for perception.

2.5 ALL-BASED DIPOLE SOURCE DETECTION

The localization ability of underwater objects with the help of ALL system can effectively improve the viability of robotic fish in underwater environment. In addition to the anti-Kármán vortex street, a near-dipole flow field is generated by the fin while fish swims. This can also explain how predators capture preys [95]. Dipole oscillation source detection has become a common problem in hydrodynamics and the development of ALL. While the dipoles are vibrating or moving in a certain way, the pressure and the flow speed will change accordingly. By measuring these information with the help of ALL, we can infer the motion of the object for further study. Carriers boarded with ALL mentioned in this section are shown in Figure 2.11.

Tang *et al.* represented an array of 8 pressure sensors installed on the surface of an underwater vehicle (Figure 2.11(a)) inspired by lateral line for near-field detection in 2019. The pressure field generated by vibrating sphere which was simulated as an underwater pressure source was derived by means of linearizing the kinematic and dynamic conditions of the free surface wave equation. The fish-shaped structure they used was boarded with 8 pressure sensors. The pressure field detected by the ALL was fit with simulation results [96].

In 2007, Yang *et al.* used an array of AHCs (Figure 2.11(b)) whose sensory unit has been introduced in section 2.2 (Figure 2.5(b)) to locate and track the dipole source. As for mapping the pressure field produced by the dipole source, the array performed well and the results was consistent in experiments and theory. As for tracking the trail, a cylinder was put in a steady flow with the speed of 0.2 m/s to simulate the hydrodynamic trail which was dominated by Kármán vortex street consequently [66].

In 2010, the same group developed an ALL by wrapping 15 pressure sensor around a cylinder (Figure 2.11(c)) to mimic real fish and proved its localization capability. They used a beamforming algorithm to image hydrodynamic events in a 3-D domain. Consequently, the ALL sensors was demonstrated to be able to localize a dipole source and a tail-flicking crayfish accurately in varieties of conditions [7].

Additionally, Asadnia *et al.* packaged $Pb(Zr_{0.52}Ti_{0.48})O_3$ thin-film piezoelectric pressure sensors for underwater sensing in 2013. The array

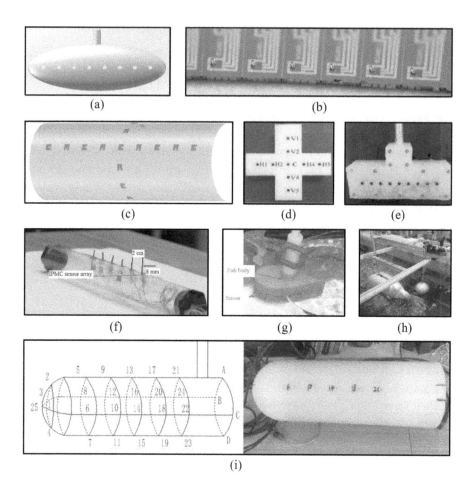

Figure 2.11 Different carriers boarded with ALL mentioned in section 2.5. (a) The fish-shaped prototype inspired by the trout lateral line in [96]. (b) Photo of an AHC sensor array used in [66]. (c) Diagram of the ALL showing the biomimetic neuromast layout in [7]. (d) The design of ALL in [10]. (e) The design of ALL in [97]. (f) An experimental prototype of IPMC-based lateral line system used in [98]. (g) A prototype of an ALL with sensors used in [99]. (h) Robotic fish with the PVDF sensor along with the oscillating sphere in [20]. (i) Sensor layout and lateral line carrier physical map in [100].

of 2 by 5 sensor has been showed in section 2.2 (Figure 2.6(a)). While the dipole was driven at the frequency of 15 Hz and moved parallel to the array, by measuring the maximum peak-to-peak output of the sensors, they approximately estimated the position of the dipole source. To detect the flow velocity generated by the dipole source, the array showed a resolution of 3 mm/s [33].

Abdulsadda *et al.* put forward an array of ALL sensors based on the sensing capability of IPMC in 2011. The signals were processed through a widely-used neural network, which was similar to the biological counterpart. The ALL has been presented in the section 2.2 as well (Figure 2.6(c)). Experiments proved that the ALL could effectively locate the dipole source and the flapping tail. Moreover, the more sensors was used, the more precise the results were [69]. In 2013, based on an analytical model of flow field produced by the dipole source, they presented another two schemes, Gauss Newton (GN) algorithms which were used to solve the nonlinear estimation problem by means of linear iteration and Newton Raphson (NR) algorithms which were used to solve the nonlinear equation under the condition of first-order optimality, to locate the dipole source and estimate amplitude and direction of the vibration. Additionally, they improved the design of intra-sensor spacing (Figure 2.11(f)) of the lateral line by analysis based on Cramer-Rao bound (CRB). With 19 dipole sources placed along an ellipsoidal track, the simulation and experiment results both proved the accuracy of this model [98, 101, 102]. In order to reduce the influence of uncertainty in measurements and flow model and thus identify a vibrating dipole accurately, Tan *et al.* developed a specialized bi-level optimization methodology to optimize the design parameters of the ALL [99] (Figure 2.11(g)).

In the aspect of using hot-wire flow sensors to detect the dipole source, Pandya *et al.* made great contributions in 2006. Section 2.2 has introduced the hot-wire ALL developed by them (Figure 2.9(a)). To make full use of the ALL, they reported on the implementation of a algorithm consisting of template training approach and the modeling approach based on minimum mean-squared error (MMSE) algorithm in order to locate and track a vibrational dipole source [78].

In 2010, by detecting the parallel and the perpendicular velocity components, Dagamseh *et al.* used an array of hair flow sensors to reconstruct the velocity field induced by the dipole source in the air and measure the distance [103–106]. In 2013, they employed beamforming techniques and improved the performance of the sensor array [107].

Inspired by the sensory abilities of lateral line, Zheng *et al.* developed an ALL composed of 9 underwater pressure sensors forming a cross (Figure 2.11(d)) to locate a dipole source in 2018. For the sake of handling nonlinear pattern identification problem, they adopted the method of generalized regression neural network which performed well on condition that the array is below 13 cm away from the dipole source [10]. Besides, the same group develop another ALL composed of 9 pressure sensors in a straight line (Figure 2.11(e)) to locate the dipole source. Lin *et al.* modeled the pressure field of the dipole source and obtained the position through least square method by means of the information acquired by the pressure sensors [97]. Ji *et al.* employed the same ALL and put forward a new method named MUSIC (multiple signal classification) for the purpose of locating dipole source with high-resolution based on spatial spectrum estimation. Moreover, they also presented a MVDR (minimum variance distortionless response) method which improved the previous Capon's method (an adaptive beamforming-based method). For further study, MUSIC method showed a potential to locate two close dipole sources [108]. In 2019, Ji *et al.* established a quantitive and method-independent Cramer-Rao lower bound (CRLB) model to evaluate the localization performance of the above two methods, least square and MUSIC, which provided guidance on the optimal design of the ALL with the least sensors and the most appropriate spacing [109].

In 2018, Liu *et al.* developed an ALL consisting of 25 high precision pressure sensors (Figure 2.11(i)). They established a mathematical model of the dipole source through Euler equation and plane potential flow theory to describe the relationship between the characteristic parameters of source and the surface pressure of the underwater vehicle and successfully detected the position, frequency and amplitude of the source through a neural network model. The consistency of simulation and experiment results demonstrated the effectiveness of the method [100].

In 2018, Yen *et al.* adopted the potential flow theory to predict the hydrodynamic pressure and presented a method to follow periodic stimuli generated by an oscillating source. The fin of the robot was regarded as an oscillator. By subtracting the pressure induced by the robotic fish from the pressure measured by the PVDF sensor (Figure 2.11(h)), they acquired the pressure generated by the source. Based on this, the robotic fish was able to adjust the amplitude, frequency, offset according to the phase difference [20]. Furthermore, this method lays a solid foundation for controlling the robotic fish to swim in a school.

In 2019, Wolf *et al.* used a 2-D array of 8 all-optical sensors to measure the velocity profiles of a underwater object and then adopted feedforward neural network and recurrent neural network to reconstruct the position of the object [110, 111]. Furthermore, they implemented near field object classification based on hydrodynamic information with an Extreme Learning Machine neural network. This method provided more information about the shape comparing to other 2-D sensing array [112].

In this section, we have put emphasis on discussing applications of ALL in locating and tracking the dipole source, especially the difference in approaches to reaching the results. Table 2.4 as follows can be a summary of this section. Flow induced by oscillating sources is only one of the most basic forms of water flows and similar to wake flow generated by real fish. Based on the results above, we can also conduct natural environment experiments in which ALL is installed on robotic fish to follow real fish, even fish school and locate them in real time.

2.6 FLOW-AIDED CONTROL OF UNDERWATER ROBOTS USING ALL SYSTEM

Lateral line plays an important part in sensing flow for fish school, which has inspired scientists worldwide to devote to developing underwater vehicles boarded with ALL system consisting of an array of sensors. The precious sections mainly presented applications based on a static ALL. If the ALL is moving like fish, the sensing difficulty will greatly increase. Assisted by ALL sensors, underwater vehicles are in a position to obtain fluid information accurately and effectively, providing the possibility to implement vehicles motion pattern identification and autonomous control. Varieties of experiments have been carried out in this domain, such as pattern identification, motion parameters (speed and direction) estimation and control, localization, obstacles detection and avoidance, energy consumption reduction and neighborhood robotic fish perception. Carriers boarded with ALL mentioned in this section are shown in Figure 2.12.

In 2013, Akanyeti *et al.* firstly dived into the problem of hydrodynamic sensing on condition that the ALL is moving. Based on Bernoulli equation, they presented a formula about pressure detected by the ALL and the moving velocity and acceleration, which laid a solid foundation for the following study [116]. Additionally, Chambers *et al.* made a study of using a vertically or horizontally moving ALL to sense local flow. They came to a conclusion that a moving ALL performed

TABLE 2.4 Classification of Existing Studies in Dipole Source Detection

Author	ALL sensors	Approaches
Tang *et al.* 2019 [96]	8 eight pressure sensors	Linearizing the kinematic and dynamic conditions of the free surface wave equation
Yang *et al.* 2007 [66]	An array of AHC sensors	Measuring the maximum peak-to-peak out put of the sensors
Yang *et al.* 2010 [7]	15 biomimetic neuromasts	Beamforming algorithm
Pandya *et al.* 2006 [78]	An array of 16 hot-wire anemometers	Template training approach and the modeling approach
Asadnia *et al.* 2013 [33]	An array of 2 by 5 pressure sensors	Maximum pressure signal
Abdulsadda *et al.* 2011 [69]	5 IPMC sensors	Neural network
Abdulsadda *et al.* 2013 [98] Abdulsadda *et al.* 2012 [101,102]	6 IPMC sensors	Gauss Newton and Newton Raphson algorithms
Ali Ahrari *et al.* 2016 [99]	Multiple flow velocity sensors	Bi-level optimization methodology
Zheng *et al.* 2018 [10]	9 underwater pressure sensors forming a cross	Generalized regression neural network
Lin *et al.* 2018 [97]	9 pressure sensors in a straight line	Least square method
Ji *et al.* 2018 [108]		MUSIC, MVDR
Ji *et al.* 2019 [109]		CRLB model
Liu *et al.* 2018 [100]	25 high precision pressure sensors	Euler equation, plane potential flow theory, and neural network
Yen *et al.* 2018 [20]	A PVDF sensor	Potential flow theory
Dagamseh *et al.* 2009 [103,104] Dagamseh *et al.* 2010 [105,106] Dagamseh *et al.* 2013 [107]	An array of hair flow-sensors	Detecting the parallel and the perpendicular velocity components, beamforming techniques
Wolf *et al.* 2019 [110,111] Wolf *et al.* 2020 [112]	8 all-optical flow sensors	Feed-forward neural network and recurrent neural network

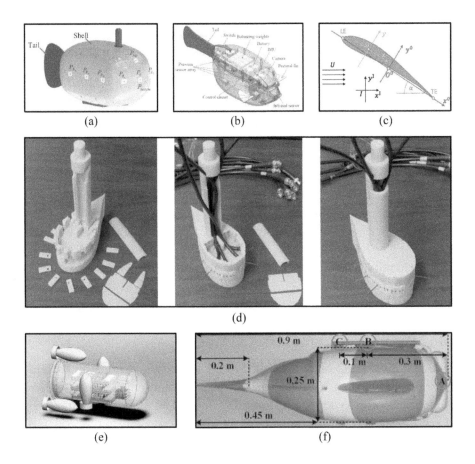

Figure 2.12 Different carriers boarded with ALL mentioned in this section 2.6. (a) The robotic fish prototype in [37]. (b) The mechanical structure and electronics of the robotic fish used in [113]. (c) Illustration of reference frames I and O. TE denotes the trailing edge and LE the leading edge in [114]. (d) Modular design of robotic foil in [19]. An array of eight IPMC sensors are installed below an array of pressure sensors. (e) Conceptual 3-D drawing of the vehicle in [115]. (f) Photographs of the robotic fish in [21].

better than static [12]. For further studies, a moving carrier boarded with ALL which has a great sensing ability can provide more valuable experimental data. The project FILOSE using a robotic fish showed in Figure 2.10(b) not only aimed to study how fish perceive and respond to fluid simulation, but also constructed a bioinspired robot on the basis of it. Boarded with ALL sensors, robotic fish measured the data from surrounding fluid environment which provided information on hydrodynamic features. And then by analysis, the relationship between fluid data and kinestate of robotic fish was established, showing a new idea of robotic fish self-control. Several experiments have been conducted as follows: 1) detecting direction while swimming against the flow, 2) swimming along a predetermined trajectory, 3) maintaining stable position in constant current, 4) reducing energy consumption in turbulence, 5) reducing energy consumption by maintaining a stable position in the hydrodynamic shadow, and 6) conducting control-experiments between real fish and robotic fish [117]. In this section, we will introduce flow-aided control of underwater robots using ALL system by category.

2.6.1 Pattern Identification

In 2020, Zheng *et al.* made a breakthrough in motion parameters estimation of a robotic fish boarded with 11 pressure sensors (MS5803-01BA) (Figure 2.12(a)). When the robotic fish moved at a specific state, such as rectilinear motion, turning motion, gliding motion, and spiral motion, they established a model combining the motion parameter including linear velocity, angular velocity, motion radius, etc. and the superficial hydrodynamic pressure variations. Robotic fish acquired the motion parameters based on the pressure detected by the ALL and then predicted the trajectory [118]. This work is of great importance for future study on self-trajectory-control.

In 2014, Liu *et al.* conducted experiments that robotic fish (Figure 2.12(b)) sense pressure information while swimming in different gaits such as forward swimming, turning, ascending, and diving. And then based on feature points extracted from the data, they adopted a subtractive clustering algorithm to recognize the swimming gaits of robotic fish [119]. The success of this approach lays a solid foundation for quick control of robotic fish.

2.6.2 Motion Parameters (Speed and Direction) Estimation and Control

In regard to the control of robotic fish, there have been a large number of results as well. Instantly, Kruusmaa *et al.* implemented rheotaxis behavior in robotic fish in 2011. With the pressure sensors detecting flow information, they put forward a linear control law which helped the robotic fish to adjust the beat frequency in order to maintain position in the steady flow [120]. In 2013, they used a 50cm-long robotic fish (Figure 2.10(b)) mimicking the geometry and swimming mode of a rainbow trout and put forward a formula for estimating the speed. The experiments have proved the validity of it [85]. Inspired by the Braitenberg vehicle 2b, they installed two pressure sensors on both sides of the head which could detect the pressure difference on the left and right sides of the robotic fish [121]. And then it was able to change the direction of swimming according to the difference to remain stable. Ulteriorly, they realized the position estimation and position stability of the robotic fish in steady water flows and behind solid objects [85].

Additionally, Xie *et al.* designed a robotic fish inspired by the geometry and swimming pattern of an ostraciiform boxfish shown in Figure 2.12(b). The robotic fish is boarded with an ALL composed of an array of 11 pressure sensors (Consensic CPS131) and an inertial measurement unit (IMU). The former is used for fluid dynamic pressure data acquisition while the latter is used to monitor the robot pitch, yaw and roll angles. In order to reduce errors on account of sensors' inaccuracy and instability, they employed an optimal information fusion decentralized filter in 2015, as a consequence of which, the accuracy of speed estimation has been greatly improved. The speed estimation formula is derived from Bernoulli principle and corrected by local filter from ALL and IMU [113]. Furthermore, in 2016, they put forward a nonlinear prediction model including distributed pressure and angular velocity to estimate the speed of robotic fish [122].

Moreover, Paley *et al.* also presented something new in flow speed and angle-of-attack detecting in 2015. They used a new type of flexible robotic fish whose shape is Joukowski-airfoil (Figure 2.12(c)) with distributed pressure sensors. Flow speed, angle-of-attack, and foil surfaces were estimated by a recursive Bayesian filter assimilation pressure measurement. And they combined an inverse-mapping feedforward controller based on an average model derived for periodic actuation of angle-of-attack and a proportional-integral feedback controller utilizing the estimated flow information to implement the closed-loop speed-control strategy [19, 114, 123].

2.6.3 Obstacles Detection and Avoidance

Furthermore, extensive efforts have been done in obstacle detection and avoidance. Inspired by sensitivity of lateral line to the presence of oncoming currents and walls or obstacles, Paley *et al.* developed a kind of wing underwater vehicle boarded with ALL sensors in 2015 (Figure 2.12(d)). Employing potential flow theory, they simulated the flow field around vehicle in the situation that the flow was uniform and there were obstacle upstream. A nonlinear estimation model of free stream flow speed, attack of angle and relative position of obstacles by measuring local flow speed and pressure difference was derived theoretically. In order to implement the stability of swimming direction and position behind obstacles, they presented a recursive Bayesian filter. Finally, they discussed the Closed-loop control strategy ulteriorly [19].

In addition, Martiny *et al.* studied intensively in obstacles detection and avoidance in 2009. They developed an autonomous underwater vehicle equipped with 4 ALL sensors (Figure 2.12(e)). Using hot-wire anemometry, it measured local flow speed around the vehicle, which was proved related to the distance between obstacles and the vehicle theoretically and experimentally [115].

Yen *et al.* have made a breakthrough in obstacles detection and navigation in 2017. They used a robotic fish boarded with 3 ALL sensors (Figure 2.12(f)) to measure nearby pressure variations, on the basis of which, they presented a way to control a robotic fish to swim along a straight wall. In theory, the tail of the robotic fish was regarded as an oscillating dipole in a 2-D potential flow approximately and the wall effect was described by an image dipole on the opposite side of the wall. The robotic fish responded to the pressure variations in order to keep a fixed distance from the wall. A qualitative relationship between velocity and wall effect was concluded in this research [21].

2.6.4 Neighborhood Robotic Fish Perception

Not only have there been various results with respect to a single robotic fish, large amounts of experiments have also been carried out in multibody control. Lots of work in this research field has been done by Xie *et al.* In 2015, they came to the conclusion by experiments that robotic fish (Figure 2.12(b)) was capable of sensing the beating frequency of the robot swimming in front and the distance between the two robots with the help of ALL system [124]. In 2017, they used the ALL system to detect the reverse Kármán-vortex street-like vortex wake generated by its adjacent robotic fish. By extracting meaningful information from the pressure variations caused by the reverse Kármán-vortex street-like vortex wake, the oscillating frequency/amplitude/offset of the adjacent robotic fish, the relative vertical distance and the relative yaw/pitch/roll angle between the robotic fish and its neighbor were sensed efficiently [35]. This progress lays a solid foundation for multi-body interactions research in the future.

In 2019, Zheng *et al.* used a robotic fish which is shown in Figure 2.12(a) to conduct neighborhood perception experiments. Firstly, based on Bernoulli principles, they established a theoretical model to describe the hydrodynamic pressure variations on the surface of two adjacent robotic fish which swam diagonally ahead another [37]. Then, they utilized dye injection technique, hydrogen bubble technique, and computational fluid dynamics simulation to study the vortices induced by the robotic fish separation. Besides, on the basis of the previous model, they also presented the relationship describing longitudinal separations and superficial hydrodynamic pressure variations of two robotic fish [36]. This progress provided a new method for robotic fish school sensing.

In addition to the results mentioned above, some other applications of ALL in robotic fish control are introduced below. With respect to localization, Muhammad *et al.* produced preliminary results by flow feature extraction and comparison of compact flow features, based on which they developed an underwater landmark recognition technique in 2015 [125]. This technique enables robotic fish to recognize locations that it has previously visited both in semi-natural and natural environments. In 2017, Francisco *et al.* proposed a map-based localization technique that employd simulated hydrodynamic maps. They used a computational fluid dynamics model to generate a flow rate diagram. Hydrodynamic information was acquired by ALL systems and analyzed to estimate the speed. Compared with the flow rate diagram, the location of the robotic fish was found out in the flow rate maps which was simulated from hydrodynamic results [126].

As for energy consumption, Kruusmaa *et al.* conducted a control experiment that the robotic fish swims in steady flow, behind a cylinder and behind a cuboid in 2013. Consequently, swimming behind a cuboid consumed the least energy. They assumed that this phenomenon was on account of the presence of the well-defined suction zone behind the cylinder. Both obstacles avoidance and energy consumption reduction are of giant significance to the navigation of robotic fish [85].

In this section, we have focused on the application of ALL system in robotic fish control. Table 2.5 as follows lists different projects mentioned above and related ongoing studies. Similar to the previous sections, robotic fish in this section is mostly static or move in a simple state, such as rectilinear motion and turning motion in laboratory environment. However, the motion of real fish and the real underwater environment are much more complicated, which makes it more difficult for underwater sensing. To solve these problems, we need to improve the sensing system and establish new control algorithms for natural environment studies. Additionally, perception of underwater obstacles provides a new approach for underwater environment reconstruction and results of neighborhood robotic fish perception can be a basis of control of multi robotic fish, both of which are potential to promote underwater exploration.

2.7 DISCUSSION

In the previous sections, we have reviewed ALL sensors based on different principles and applications in flow field characteristics identification, dipole source detection and control of underwater robots. Although fish lateral line provides inspiration for the design of ALL sensors and underwater detection, it also serves as a strict standard for research. There have been great progress in this area, however, performance of existing ALL system is still quite far from that of real fish.

ALL sensors which have been developed are no match for that of real fish through evolution in sensitivity, stability, coordination, and information processing. We can optimize the design of the sensitive element and consider the resonance frequency for a better perception of ALL sensors. As for the stability, measurement errors in different temperature or pressure conditions should be taken into account. Additionally, waterproofing measures is necessary for the normal operation of sensors in the harsh conditions underwater. The development of new materials and micromachining technology provides possible methods for improvements in both areas. Not only should the performance of a single sensory unit be improved, the coordination of an array of ALL sensors is also important. Existing arrays of ALL sensors are mainly composed of a single type of sensors (pressure sensors or flow sensors) and arranged regularly, which is quite from that of real fish. Real lateral line consists of SNs and CNs for a comprehensive perception of surrounding environment and sensing cells are distributed in a specific pattern for a better sensing. Using pressure sensors and flow sensors simultaneously is a potential method to optimize the array of ALL sensors. Besides, we can put forward evaluation indexes of ALL in order to find out the best placement of sensors on the surface of underwater robots. In terms of information processing, the fish lateral line has many different sensory functions, which can be a reference standard for ALL sensors. Many algorithms in velocity measurements and dipole source detection have been put forward, but the sensory ability of ALL does not stop here. ALL has the potential to sense the obstacles, fish schools and even reconstruct the surrounding water environment, which is a basis of follow-up research on control of underwater robots.

As for hydrodynamic characteristics identification, there have been many results based on ALL system. However, existing results are mainly based on laboratory experiments where the motion of ALL carrier is simple such as static state and rectilinear motion and the water environment is stable. By contrast, the motion of real fish and underwater environment are much more complicated, which will greatly increase the difficulty of experiments. With the development of ALL sensors, natural experiments are necessary for complete imitation of real fish lateral line.

In dipole source detection, many methods and algorithms have been presented because oscillatory flow is one of the most basic flows and can be used to simulate the wake produced by the wagging tail of fish, which is important for further study on location and track of real fish. But the matching degree of oscillating flow and fish wake needs further exploration. On the basis of it, more experiments on real-time location of fish school and mimicking the group behavior of fish may become a focus in the future.

Flow-aided control of underwater robots is a promising project which has a wide range of applications in marine exploration. The first priority is to establish the motion model of underwater robots based on hydrodynamics theory or date-driven methods which provides a way for self-identification of motion pattern in the conditions where it is impossible to observe. Establishment of model can be a theoretical instruction for control strategies. In addition, obstacles recognition and avoidance is another necessary ability for autonomous control. Based on the location of obstacles, robots is potential to reconstruct the surrounding environment and plan an optimal path for navigation. Except for a single robotic fish, perception and control of a robotic fish group may greatly improve the efficiency of underwater exploration and the success rate of underwater missions with the help of ALL.

2.8 CONCLUSION

We briefly introduced the morphology and mechanism of the lateral line for a further understanding. And then we put emphasis on discussing progress in biomimetics inspired by lateral line. Different kinds of sensors based on different principles have been explored, which provides a new approach for fluid information detection. Furthermore, scientists have arranged sensors to fabricate ALL system on underwater vehicle bodies to assist in underwater detection, which is superior to traditional methods in applicability and flexibility. For hydrodynamic environment sensing and characteristic detection, we have made great progress in flow field characteristics identification, flow velocity and direction detection, vortex street properties detection. Additionally, various algorithms have been put forward to locate the dipole source. Recently, ALL system have been used for the autonomous control of underwater robotic fish.

In future, based on existing results, we need to develop sensor units demonstrating higher sensitivity and optimize the distribution of ALL on the robotic fish surface. Combining a high-performance sensor array with a more efficient signal processing algorithm, robotic fish is potential to show a sensory ability close to real fish, which is a foundation for natural environment experiments in the following studies. It is necessary to present new universal control algorithms for the autonomous control of robotic fish based on ALL systems. Perception of underwater obstacles provides a new approach for underwater environment reconstruction, which is important for underwater location and navigation. Moreover, application of ALL in the control of robotic fish school will also become a focus issue, which is important in multi-robot interaction and cooperation for future ocean exploration. With the development of new materials, fabrication technology and artificial intelligence, the performance of ALL system will be comparable to that of real fish.

The development of ALL systems provides a powerful tool for ocean exploration. Although great efforts have been devoted in this area, there is a long way to go to realize the comparable sensory ability and autonomous control of robotic fish to real fish.

Boxfish-Like Robot with an Artificial Lateral Line System

TWO BOXFISH-LIKE ROBOTS with an artificial lateral line system (ALLS) were used for ALLS-based sensing experiments presented in the following chapters. The robots include one multi-fin-actuated boxfish-like robot and one caudal-fin-actuated boxfish-like robot. The former one was inspired by a species of cube boxfish (*Ostracion cubicus* [127], Figure 3.1). The boxfish has a prismatic and not streamlined shell with chiseled spine structures at the edge of the shell. In addition, the outer surfaces between adjacent spines have concave and convex extents. Such a unique outer structure causes the vortex separation downstream, which influences the maneuverability and stability of boxfish. However, it is still not clear whether, and if so how, such a unique shell improves or reduces the boxfish's motion performance. As a result, to suppress the shell-induced influence on motions of the fish robot, we simplified the shell and designed the other caudal-fin-actuated fish robot with a cuboid-shaped shell. Based on the two boxfish-like robots, we conducted online state estimation of a boxfish-like robot using artificial lateral line system and artificial lateral line-based local sensing between two adjacent boxfish-like robots.

DOI: 10.1201/b23027-3

Figure 3.1   Real cube boxfish (Ostracion cubicus [127]).

3.1   MULTI-FIN-ACTUATED BOXFISH-LIKE ROBOT

The robotic fish used is inspired by a species of cube boxfish (Ostracion cubicus [127]). Figure 3.2 shows the dimensions and hardware configurations of the robotic fish. Its size (Length × Width × Height) is about 40 cm × 14.1 cm × 13.2 cm. It mainly consists of a sealed shell, a pair of pectoral fins and a caudal fin. The electrical system which consists of rechargeable batteries, steering engines, circuit boards, and sensors is wrapped in the sealed shell. Specifically, the paired pectoral fins and the caudal fin are respectively connected with three steering engines for producing propulsive forces.

As shown in Figure 3.3, the sensors include an inertial measurement unit (IMU), a camera, an infrared sensor, and nine pressure sensors. IMU, camera, and infrared sensor are all placed at the central area in the front of the robot. IMU provides the acceleration and attitude information of the robotic fish so that its three-dimensional motions can be monitored. Camera and infrared sensor are used together to capture the surrounding environment. Thus the robotic fish can avoid obstacles and locate itself. Pressure sensors are distributed over the surface of the shell to establish an ALLS, for measuring the external pressures. Multiple 32-bit micro controllers (STM32F405 and STM32F103) on the circuit boards serve the functions of sensor data acquisition and steering engine control. The bottom circuit board is used for controlling steering engines and collecting data of camera, infrared sensor and IMU. In addition, a credit card-sized micro-computer called Raspberry Pi is adopted as a main processor of the robotic fish. Based on a Linux system

(Detain) installed on Raspberry Pi, the robotic fish can be operated autonomously.

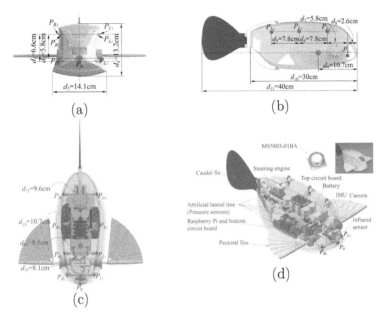

Figure 3.2 Dimensions and hardware configurations of the robotic fish. (a) Front view. (b) Side view. (c) Top view. (d) Isometric view. The red points in (a), (b), and (c) indicate the centers of mass of the robotic fish.

A central pattern generator (CPG) based controller is designed to control the locomotion of the robotic fish [128]. Specifically, through controlling the steering engines connected with the paired pectoral fins and the caudal fin with given CPG parameters, the fins are able to oscillate with given frequencies, amplitudes, and offsets. Using the propulsive forces generated by the fins, the robotic fish is able to realize multiple three-dimensional swimming patterns which specifically include forward/backward swimming, left/right turning, upward/downward swimming, and clockwise/counter-clockwise yawing/pitching/rolling. More details about the locomotion of the robotic fish can be found in [128,129]. Such multiple swimming patterns enable the robotic fish to form multiple relative positions and attitudes to its neighbor. As shown in Figure 3.4, the relative positions can be classified into relative lateral position, relative longitudinal position and relative vertical position. The relative attitudes include relative yawing, relative pitching and relative rolling.

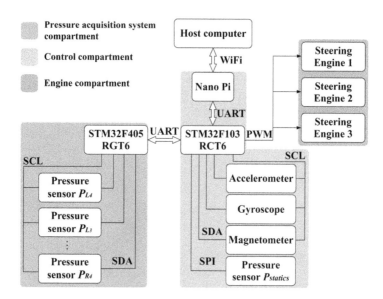

Figure 3.3 The diagrammatic sketch of the electrical system of the boxfish-like robot.

With the use of lateral line, fish can sense the velocity of the flow and the hydrodynamics pressure in the flow field. Considering that we have mainly focused on investigating the hydrodynamic pressure variations existed in underwater environment in the experiments, thus we have adopted a kind of commercial pressure sensor MS5803-01BA from TE Connectivity Ltd. to establish the ALLS. The pressure sensor provides a resolution of 1.2 Pa, thus making it have the capability to measure the tiny HPVs in the water. Besides, the size (Length × Width) of the pressure sensor is about 6.2 mm × 6.4 mm. It is so small that we can mount multiple pressure sensors in an array, imitating boxfish's lateral line [130]. As shown in Figure 3.2, the ALLS consists of nine pressure sensors which are marked as P_0, P_{L_1}, P_{L_2}, P_{L_3}, P_{L_4}, P_{R_1}, P_{R_2}, P_{R_3}, and P_{R_4}, respectively.

3.2 CAUDAL-FIN-ACTUATED BOXFISH-LIKE ROBOT

Figure 3.5(a) and (b) shows the dimensions and hardware configurations of the robotic fish. Its size (Length × Width × Height) is about 29.1 cm × 11.6 cm × 13.4 cm. It mainly consists of a 3D-printed shell, an engine compartment, a control compartment, a battery compartment, a pressure acquisition system compartment, and a tail. Using lateral line,

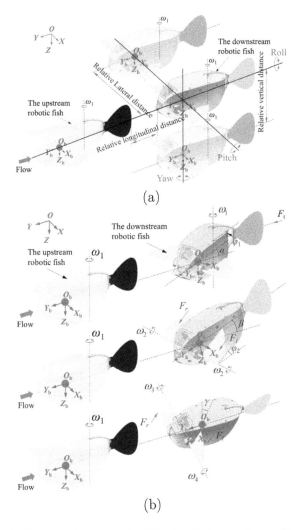

Figure 3.4 Relative positions and attitudes between two adjacent robotic fish. (a) Relative positions between two adjacent robotic fish and definition of the attitude motions. The red point indicates the center of mass of the robotic fish. (b) Relative attitudes between two adjacent robotic fish. The relative attitudes include relative yaw angle α, relative pitch angle β, and relative roll angle γ. ω_1, ω_2, ω_3, and ω_4 indicate the angular velocities of the fins. φ_1 and φ_2 indicate the offsets of the fins. $OXYZ$ and $O_bX_bY_bZ_b$ indicate the global inertial coordinate system and the fish body-fixed coordinate system, respectively. F_t, F_l, and F_r indicate the propulsive force generated by the oscillating caudal fin, the left pectoral fin, and the right pectoral fin, respectively.

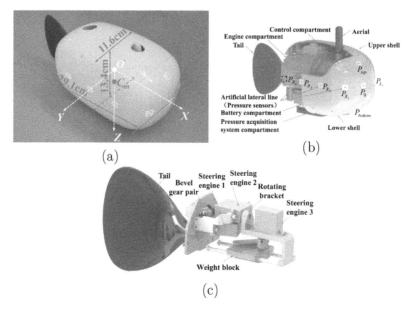

(a)

(b)

(c)

Figure 3.5 Hardware configurations of the robotic fish. (a) The robotic fish prototype. $OXYZ$ indicates the body-fixed coordinate system of the robotic fish. The origin O is fixed at the intersection of horizontal section and longitudinal section of the robotic fish, above center of mass C_m of the robotic fish. (b) CAD model of the robotic fish. (c) The diagrammatic sketch of the interior of the engine compartment.

fish can sense the flow velocity and pressure in underwater environment. Considering that we have mainly focused on investigating the PVs in the experiments, we have adopted a kind of commercial pressure sensor MS5803-01BA from TE Connectivity Ltd. to establish the ALLS. The pressure sensor provides a resolution of 1.2 Pa, which makes it have the capability of measuring tiny PVs in the water. The sensor range of the currently used pressure sensors is from 1.0×10^3 Pa to 1.3×10^5 Pa. Considering that the standard atmospheric pressure is about 1.01×10^5 Pa, so the pressure sensor can be used in a depth of at most 2.9 m. There are other types of pressure sensors which have larger ranges. We can also acquire those pressure sensors according to practical applications. The size (Length × Width) of the pressure sensor is about 6.2 mm × 6.4 mm. It is so small that we can mount multiple pressure sensors in an array. As shown in Figure 3.5(b), eleven pressure sensors named P_{top}, P_{bottom}, P_0, P_{Lm}, and P_{Rm} ($m = 1, 2, 3, 4$) are distributed over the surface of the shell to establish an ALLS, for measuring the PVs around fish body.

Figure 3.5(c) shows the interior of the engine compartment. Three steering engines which serve different functions are wrapped in the engine compartment. Specifically, steering engine 1 is connected with the tail, for producing propulsive force. Steering engine 2 is used to drive a rotating bracket to which steering engine 3 and a link mechanism are connected. Steering engine 3 is used for driving the above-mentioned link mechanism to which a weight block is connected. Through controlling steering engine 2 and 3, the weight block is able to move along the direction parallel to the OX axis of the robotic fish and rotate about output shaft of steering engine 2. Through controlling the three steering engines with given frequency, amplitude and offset parameters, the robotic fish is able to realize multiple three-dimensional swimming patterns including rectilinear motion, turning motion, gliding motion, and spiral motion, as shown in Figure 3.6.

As shown in Figure 3.7, the control compartment contains an attitude and heading reference system (AHRS), several circuit boards, a credit card-sized micro-computer called NanoPi, and a pressure sensor which is named $P_{statics}$ and locates at the bottom of the compartment. AHRS consists of a triaxial accelerometer, a gyroscope and a magnetometer. It outputs yaw angle, pitch angle, roll angle, angular velocities, and accelerations of the robotic fish with a sampling rate of 50 Hz. Multiple 32-bit micro controllers on the above-mentioned circuit boards serve the functions of AHRS data acquisition and steering engines control. The pressure sensor $P_{statics}$ (MS5803-14BA, TE Connectivity Ltd.) is used for measuring the static pressure when the robotic fish is beneath the water. The static PVs can be used to calculate the depth variations of the robotic fish. NanoPi is adopted as a main processor of the robotic fish. Based on a Linux system installed on NanoPi, the robotic fish can be operated autonomously. The pressure acquisition system compartment contains a circuit board which serves the functions of collecting data of the ALLS via inter-integrated circuit (I^2C) bus and then transferring the data to NanoPi. Finally, the data were transmitted to host computer through wireless serial communication modules, with a rate of 50 Hz.

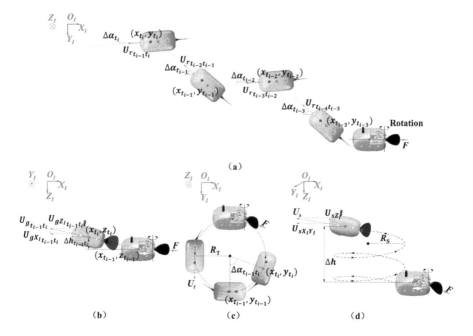

Figure 3.6 Multiple three-dimensional swimming patterns of the robotic fish. (a) Rectilinear motion. (b) Gliding motion. (c) Turning motion. (d) Spiral motion. The tails in (c) and (d) have nonzero offsets compared to those in (a) and (b). The weight blocks in (b) and (d) leave their original positions compared to those in (a) and (c). The white arrows in (b) and (d) indicate the movement direction of the weight block. $O_I X_I Y_I Z_I$ indicates the global inertial coordinate system. F indicates the tailed-generated propulsive force. $(x_{t_i}, y_{t_i}, z_{t_i})$ indicates the coordinate of the robotic fish at time $t_i(i = 1, 2, 3, ...)$ in $O_I X_I Y_I Z_I$. $U_k(k = r, t, g, s)$ indicates the movement velocity of the robotic fish. $U_{k_{t_{i-1}t_i}}(k = r, t, g, s)$ indicates the movement velocity of the robotic fish at time interval (t_{i-1}, t_i). $U_{gX_{I_{t_{i-1}t_i}}}$ and $U_{gZ_{I_{t_{i-1}t_i}}}$ indicate the velocity component along the axis $O_I X_I$ and the axis $O_I Z_I$ at time interval (t_{i-1}, t_i). $U_{sX_I Y_I}$ and U_{sZ_I} indicate the velocity component on the plane $O_I X_I Y_I$ and along the axis $O_I Z_I$. R_T and R_S indicates the turning radius and spiral radius, respectively. Δh indicates the depth variation of the robotic fish. $\Delta \alpha_{t_i}$ in (a) is the difference between the yaw angle α_{t_i} at time t and the initial yaw angle α_{t_0} of the robotic fish ($\Delta \alpha_{t_i} = \alpha_{t_i} - \alpha_{t_0}$). $\Delta \alpha_{t_{i-1}t_i}$ in (c) is the difference between the yaw angle α_{t_i} at time t_i and $\alpha_{t_{i-1}}$ at time t_{i-1} ($\Delta \alpha_{t_{i-1}t_i} = \alpha_{t_i} - \alpha_{t_{i-1}}$).

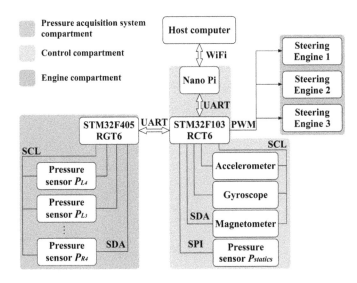

Figure 3.7 The diagrammatic sketch of control system of the robotic fish.

CHAPTER 4

Online State Estimation of a Boxfish-Like Robot Using Artificial Lateral Line System

L ATERAL LINE SYSTEM (LLS) is a flow-responsive organ system with which fish can effectively sense the surrounding flow field, thus serving functions in flow-aided fish behaviors. Inspired by such a biological characteristic, artificial lateral line systems (ALLSs) have been developed for promoting technological innovations of underwater robots. In this chapter, we focus on investigating state estimation of a freely-swimming robotic fish in multiple motions including rectilinear motion, turning motion, gliding motion, and spiral motion. The state refers to motion parameters including linear velocity, angular velocity, motion radius, etc., and trajectory of the robotic fish. Specifically, for each motion, a pressure variation (PV) model which links motion parameters to PVs surrounding the robotic fish is firstly built, then a linear regression analysis method is used for determining the model parameters. Based on the acquired PV model, motion parameters can be estimated by solving the PV model inversely using the PVs measured by ALLS. Finally, a trajectory estimation method is proposed for estimating trajectory of the robotic fish based on the ALLS-estimated motion parameters. The experimental results show that the robotic fish is able to estimate its

DOI: 10.1201/b23027-4

trajectory in the above-mentioned multiple motions with the aid of ALLS, with small estimation errors.

4.1 INTRODUCTION

Underwater navigation is one of the key technologies for underwater robots and vehicles. As stated by J.J. Leonard and H.F. Durrant-Whyte, the problem of robot navigation mainly involves three questions: "Where am I?", "Where am I going?", and "How should I get there?" [131]. The first one refers to localization of the robot. The second one and the third one respectively emphasize the destination and the approach of reaching the destination. These two questions generally involve multiple underwater robot technologies which include underwater communication, underwater target detection and identification, underwater robot control, etc. We mainly focus on the localization of underwater robots and vehicles, it is a basic problem which should be firstly solved for underwater navigation. The existing approaches of underwater localization are mainly on the basis of the acoustic positioning system (APS), the inertial navigation system (INS), the underwater global positioning system (GPS), and the optical positioning system (OPS) [132]. The above-mentioned positioning systems have several significant shortages, such as the following.

First, most of the positioning systems have extremely complex components. The APS system typically works based on sonar and doppler velocity log (DVL) [133]. Take sonar for example, sonar system typically consists of acoustic transducer arrays, electronic cabinet, and auxiliary equipment including power device, connecting cable, sonar dome, etc. [134]. Such complex components often result in high cost. Second, except for partial OPS (vision based position system), most of the above-mentioned systems cannot be fully integrated into small (centimeter-scale) underwater robots and vehicles, so these underwater robots and vehicles can only realize localization depending on the external sensor system, which is not online. Third, for APS system, INS system, and OPS system, they cannot serve functions in hostile environments with complex topography, magnetic-field interference, and poor light, respectively. Comparing to the above-mentioned sensor systems, ALLS is cheaper and still able to serve functions normally in the above-mentioned hostile environment. Besides, ALLS are small enough to be integrated into centimeter-scale underwater robots and vehicles. So it is worth exploring the practicability of ALLS in localization of underwater robots.

According to whether the position-related priori information is known, the problem of localization can be divided into two categories: global localization and position tracking [135]. The global localization is also called absolute localization. It refers to determining position of a robot in the case that both its initial position and any relative positions to other objects are unknown. The position tracking is also called relative localization which refers to determining the position of the relative robot to its known initial position.

On the basis of the above-mentioned analyzes, we mainly focus on investigating position tracking of underwater robot and vehicle. We study how to realize online trajectory estimation of a robotic fish in rectilinear motion, turning motion, gliding motion, and spiral motion using its onboard ALLS. Specifically, for each motion, several motion parameters including linear velocity, angular velocity, motion radius, etc., are used to characterize the motion. Considering that motions of the robotic fish result in pressure variations (PVs) around its shell, theoretical PV models which reflect the relationship between the motion parameters and the PVs are firstly built. Then multiple experiments are conducted to measure the actual PVs and the actual motion parameters when the robotic fish is under corresponding motions, and a linear regression analysis method is used for determining the PV model parameters. Based on the acquired PV models, we attempt to use the PVs measured by the ALLS to estimate the motion parameters. Using the estimated motion parameters, the trajectory of the robotic fish can be estimated according to a trajectory estimation method based on related theories of physics and mathematics. The errors between the estimated trajectory and the measured trajectory are also analyzed.

This chapter addresses the problem of online state estimation of a robotic fish, and contributes in the following four aspects.

1) Proposing complete PV models which link the motion parameters including linear velocity, motion radius, and angular velocity, etc. of a robotic fish to the PVs surrounding the robotic fish in 3-dimensional motions including rectilinear motion, turning motion, gliding motion, and spiral motion.

2) Conducting experiments with large scopes of parameter space for determination and validation of the proposed PV models.

3) Based on ALLS, realizing comprehensive online estimation of motion parameters, which is always a challenge for small (centimeter-scale) underwater robot in practical applications.

4) Proposing a novel method for online trajectory tracking of underwater robot, which is always an extremely critical problem. A more detailed account of the contributions is as follows.

Among the existing studies of applications of ALLS on underwater robots, the experiments have mainly been conducted in an excessively controlled experiments where the ALLS is fixed in flow environment or slowly towed with a simple linear motion. Only a few lateral line inspired sensing research has investigated freely motion of underwater robots, and the experiments have been conducted with a limited parameter space [22, 33, 116]. To our best knowledge, it is the first time that artificial lateral line system is used to investigate 3-dimensional motion of underwater robot. Moreover, the experiments are conducted in large-scale experiment parameter space. Specifically, in each motion mode, the robotic fish is actuated by various combinations of different oscillating parameters of tail and positions of weight block. On the other hand, there are only a few works have explored ALLS based estimation of motion parameters. Otar Akanyeti *et al.* have established a hydrodynamic model which reflects the forward velocity of a robotic fish and the pressure distribution on the surface of fish body using two-dimensional potential flow theory. Basing on the HPVs measured by ALLS, the speed of the robotic fish has been obtained by solving the established model inversely [116]. However, the robotic fish in [116] has been moved passively by a linear motor rig, without rotation of fish head which a freely-swimming robotic fish has. Significant differences do exist between the passive movement and free swimming of the robotic fish. Wang *et al.* have explored how to estimate the forward swimming velocity of a freely swimming robotic fish using its onboard ALLS [22]. However, the estimation model in [22] has been assumed by observation of kinematic data and validated by data-driven approach, without theoretical analysis. Besides, only an estimation model for rectilinear motion has been investigated. In this chapter, we have taken account of rhythmical motion of the freely-swimming fish and realized the ALLS based online estimation of more motion parameters. In addition, the ALLS based trajectory tracking method can be also extended to other species of underwater robots, so this chapter provides a novel guidance for localization of underwater robot. It demonstrates the effectiveness and great potential of artificial lateral line in improving the performances of underwater robots.

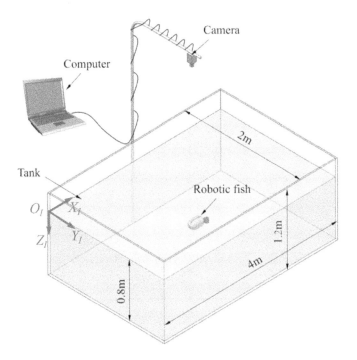

Figure 4.1 Hardware platform for the experiments. The origin O_I of $O_I X_I Y_I Z_I$ is fixed on the surface of the water and at the corner of the tank. The $O_I X_I Y_I$ plane coincides with the surface of the water while the axis $O_I Z_I$ is along the depth direction of the water.

4.2 MATERIALS AND METHODS

4.2.1 The Experimental Description

We used artificial lateral line to estimate the trajectory of the robotic fish when it was in rectilinear motion, turning motion, gliding motion, and spiral motion. As shown in Figure 4.1, the experiments were conducted in a tank whose size (Length × Width × Height) was about 4 m × 2 m × 1.2 m, and the depth of water was 0.8 m. Besides, an overhead camera with a vision tracking platform was used to capture the 2-dimensional position coordinates of the robotic fish in the tank. The attitudes of the robotic fish were monitored by the AHRS data to ensure the robotic fish was in given motions. Specifically, pitch angle of the robotic fish oscillates around 0° in rectilinear motion and turning motion, while it oscillates around a nonzero value in gliding motion and spiral motion. Yaw angle of the robotic fish oscillates around 0° in rectilinear motion

TABLE 4.1   List of Experimental Parameters

Experiments	Amplitude A (°)	Frequency f (Hz)	Offset ϕ (°)	Position of the weight block Δd (cm)
Rectilinear motion	{5, 10, ..., 30}	{1.0, 1.2, ..., 1.8, 2.0}	0	0
Turning motion	20	{1.0, 1.1, ..., 1.9, 2.0}	{20, 25, ..., 40}	0
Gliding motion	20	2.0	0	{−2.0, −1.9, −1.8, −1.75, −1.7, −1.4, −1.3, −1.2, −1.0, −0.8, −0.6, −0.4, −0.2, 0}
Spiral motion	20	3.0	20	{−1.6, −1.4, −1.2, −1.0, −0.6, −0.4, −0.2, 0}

and gliding motion, while it oscillates around a time-varying value in turning motion and spiral motion. Roll angle oscillates around 0° in the four motions. In order to satisfy the above-mentioned conditions, density of the robotic fish was well regulated to be slightly less than 1 kg/m³, and roll angle and pitch angle of the robotic fish were regulated to be around 0° according to the AHRS data when the robotic fish was located in still water, after precisely regulating the distribution of mass before experiments.

4.2.2   The Experimental Procedures

The experimental procedures were described as follows. First, locate the robotic fish in the tank. Second, record data of the pressure sensor $P_{statics}$ and the ALLS in still water for ten seconds, and the mean values of the pressure data recorded by $P_{statics}$ and ALLS were respectively denoted as $\overline{P_{s_1}}$ and $\overline{P_{alls_1}}$. Third, start the steering engine actuating the tail and the weight block with given parameters as shown in Table 4.1, for realizing given motions. A, f, and ϕ indicate the amplitude, frequency, and offset of the tail of the robotic fish. Δd indicate the distance between the weight block's current position and its initial position. When the value of Δd is positive, the weight block move toward the head. While Δd is negative, the weight block move toward the tail. Then, wait for several seconds until motion state of the robotic fish became stable. Then, record the pressure sensor $P_{statics}$ data, ALLS data, AHRS data, and position coordinates of the robotic fish for ten seconds. Values of the pressure data recorded by $P_{statics}$ and ALLS were respectively denoted as P_{s_2} and P_{alls_2}. Finally, stop the recording and then stop the robotic fish. The difference ΔP_s between P_{s_2} and $\overline{P_{s_1}}$ reflected the static PVs ($\Delta P_s = P_{s_2} - \overline{P_{s_1}}$), while the difference ΔP_{alls} between P_{alls_2} and $\overline{P_{alls_1}}$ reflected the PVs surrounding the robotic fish when it was in given motions. Specifically, $\Delta P_{alls} = P_{alls_2} - \overline{P_{alls_1}}$. Except for the recordings of pressure data, 2-dimensional position coordinates of the robotic fish are also recorded by the overhead camera when the robotic fish is moving. For each experimental parameter, the above-mentioned recordings were repeated for five times. Basing on the recorded data, the measured values of related motion parameters which include rectilinear velocity (U_r), turning velocity (U_t), spiral velocity in $O_I X_I Y_I$ plane ($U_{sX_IY_I}$), gliding velocity in $O_I X_I Y_I$ plane ($U_{gX_IY_I}$), turning angular velocity (Ω_t), spiral angular velocity (Ω_s), turning radius (R_T), and spiral radius (R_S) were calculated. Specifically, we have used the camera-recorded coordinates to calculating the distance s which the robotic fish has moved every sample time Δt, then U_r (U_t, $U_{gX_IY_I}$, and $U_{sX_IY_I}$) was the quotients of s and Δt. For R_T and R_S, they were acquired by least squares curve fitting method using the recored coordinates. Ω_t was quotients of U_t and R_T. For Ω_s, it was quotients of $U_{sX_IY_I}$ and R_S. For the depth variation Δh and velocity v_{depth} in the depth direction, they were calculated by the static PVs measured by the pressure sensor $P_{statics}$, according to (4.1) and (4.2).

$$\Delta h = \Delta P_s / (\rho g) \qquad (4.1)$$

where ΔP_s is the static PVs measured by $P_{statics}$, ρ is the density of water, and g is the acceleration of gravity.

$$v_{depth} = \Delta h / t_h \qquad (4.2)$$

where t_h is the time corresponds to Δh.

4.3 PRESSURE VARIATION MODEL

4.3.1 Theoretical Analysis for Hydrodynamic Pressure Variation

According to Lighthill's discussion in [136], the hydrodynamic pressure variations $\Delta V_{dynamics}$ on the surface of the robotic fish body can be approximated by the unsteady Bernoulli equation (4.3), basing on the assumption that the flow remains irrotational and there is no boundary layer effect.

$$\Delta V_{dynamics}(t) = -\rho \frac{\partial \Phi}{\partial t} - \frac{1}{2}\rho |\nabla \Phi|^2 \tag{4.3}$$

where $\Delta V_{dynamics}(t)$ represents the hydrodynamic pressure variation which is the difference between the pressure value at time t and the pressure value in still water. ρ is the density of water. Φ is the velocity potential.

We consider the case that the velocity potential is related with component of the translational velocity along OX axis of the body frame $OXYZ$ of the robotic fish (defined as U), pitch angle θ of fish body, and sway angular velocity ω of fish body. U, θ, and ω are all functions of time t. The velocity potential can be written as

$$\begin{aligned} \Phi &= U\Phi_U(X,Y,Z) + \theta\Phi_\theta(X,Y,Z) + \omega\Phi_\omega(X,Y,Z) \\ &= \begin{bmatrix} U,\theta,\omega \end{bmatrix} \cdot \begin{bmatrix} \Phi_U(X,Y,Z) \\ \Phi_\theta(X,Y,Z) \\ \Phi_\omega(X,Y,Z) \end{bmatrix} \end{aligned} \tag{4.4}$$

where (X,Y,Z) represents positions on the surface of the robotic fish in body-fixed coordinate system $OXYZ$. Φ_U, Φ_θ, and Φ_ω respectively represent the velocity potential associated with the movement of fish body at unit velocity, the pitching motion of fish body at unit pitch angle, and the yawing motion of fish body at unit sway angular velocity.

The time derivative of the velocity potential is expressed as

$$\begin{aligned} \frac{\partial \Phi}{\partial t} &= \frac{d}{dt} \begin{bmatrix} U,\theta,\omega \end{bmatrix} \cdot \begin{bmatrix} \Phi_U(X,Y,Z) \\ \Phi_\theta(X,Y,Z) \\ \Phi_\omega(X,Y,Z) \end{bmatrix} \\ &\quad - \begin{bmatrix} U,\theta,\omega \end{bmatrix} \cdot \begin{bmatrix} \nabla\Phi_U \\ \nabla\Phi_\theta \\ \nabla\Phi_\omega \end{bmatrix} \cdot \frac{d}{dt} \begin{bmatrix} X \\ Y \\ Z \end{bmatrix} \end{aligned} \tag{4.5}$$

where $\nabla\Phi_U = \begin{bmatrix} \frac{\partial\Phi_U}{\partial X}, \frac{\partial\Phi_U}{\partial Y}, \frac{\partial\Phi_U}{\partial Z} \end{bmatrix}$, $\nabla\Phi_\theta = \begin{bmatrix} \frac{\partial\Phi_\theta}{\partial X}, \frac{\partial\Phi_\theta}{\partial Y}, \frac{\partial\Phi_\theta}{\partial Z} \end{bmatrix}$, $\nabla\Phi_\omega = \begin{bmatrix} \frac{\partial\Phi_\omega}{\partial X}, \frac{\partial\Phi_\omega}{\partial Y}, \frac{\partial\Phi_\omega}{\partial Z} \end{bmatrix}$. $\frac{dX}{dt} = U$. $\frac{dX}{dt}$, $\frac{dY}{dt}$ and $\frac{dZ}{dt}$ indicate the velocity component along the OX axis, OY axis and OZ axis in body-fixed coordinate system $OXYZ$. Considering that $\frac{dY}{dt}$, $\frac{dZ}{dt}$, and accelerations of the robotic fish were small comparing to U, θ, and ω, so they can be neglected. That is to say, $\frac{dU}{dt} \approx 0$, $\frac{d\theta}{dt} \approx 0$, $\frac{d\omega}{dt} \approx 0$, $\frac{dY}{dt} \approx 0$, and $\frac{dZ}{dt} \approx 0$. As a consequence, (4.5) finally simplifies to

$$\frac{\partial \Phi}{\partial t} = U^2 \left(-\frac{\partial \Phi_U}{\partial X} \right) + U\theta \left(-\frac{\partial \Phi_\theta}{\partial X} \right) + U\omega \left(-\frac{\partial \Phi_\omega}{\partial X} \right) \qquad (4.6)$$

in addition,

$$\begin{aligned}
|\nabla \Phi|^2 &= |U\nabla\Phi_U + \theta\nabla\Phi_\theta + \omega\nabla\Phi_\omega|^2 \\
&= U^2(\nabla\Phi_U)^2 + \theta^2(\nabla\Phi_\theta)^2 + \omega^2(\nabla\Phi_\omega)^2 \\
&\quad + 2U\theta(\nabla\Phi_U)(\nabla\Phi_\theta) \\
&\quad + 2U\omega(\nabla\Phi_U)(\nabla\Phi_\omega) \\
&\quad + 2\theta\omega(\nabla\Phi_\theta)(\nabla\Phi_\omega)
\end{aligned} \qquad (4.7)$$

(4.3) together with (4.6) and (4.7) gives

$$\begin{aligned}
\Delta V_{dynamics}(t) = &C_1 U^2 + C_2 \theta^2 + C_3 \omega^2 + C_4 U\theta \\
&+ C_5 U\omega + C_6 \theta\omega + C_7
\end{aligned} \qquad (4.8)$$

where

$$\begin{aligned}
C_1 &= \rho\frac{\partial \Phi_U}{\partial X} - \frac{1}{2}\rho(\nabla\Phi_U)^2 \\
C_2 &= -\frac{1}{2}\rho(\nabla\Phi_\theta)^2 \\
C_3 &= -\frac{1}{2}\rho(\nabla\Phi_\omega)^2 \\
C_4 &= \rho\frac{\partial \Phi_\theta}{\partial X} - \rho(\nabla\Phi_U)(\nabla\Phi_\theta) \\
C_5 &= \rho\frac{\partial \Phi_\omega}{\partial X} - \rho(\nabla\Phi_U)(\nabla\Phi_\omega) \\
C_6 &= -\rho(\nabla\Phi_\theta)(\nabla\Phi_\omega) \\
C_7 &= constant
\end{aligned} \qquad (4.9)$$

In our previous investigations, we have found that the data recorded by the pressure sensors slightly drift with the time even at zero external velocity. Besides, the background noise may also cause the static drift of the pressure sensor. So C_7 is incorporated into $\Delta V_{dynamics}$ for compensating the above effects. Considering that Φ_U, Φ_θ, Φ_ω, $\nabla\Phi_U$, $\nabla\Phi_\theta$, and $\nabla\Phi_\omega$ are all functions of (X, Y, Z), so C_n $(n = 1, 2, 3, 4, 5, 6, 7)$ is also function of (X, Y, Z). That is to say, C_n depends on the geometry of the robotic fish. We investigate the PVs at the positions where the pressure sensors of ALLS lie. For simplifying the analysis, we replace

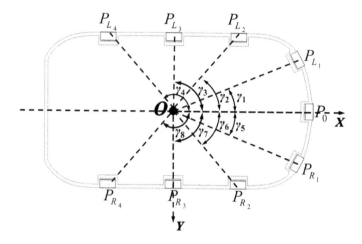

Figure 4.2 Top view of the robotic fish model and the definition of γ. $\gamma_0 = 0$, $\gamma_1 = -\gamma_5 = 0.39$ rad, $\gamma_2 = -\gamma_6 = 0.86$ rad, $\gamma_3 = -\gamma_7 = 1.57$ rad, and $\gamma_4 = -\gamma_8 = 2.28$ rad.

(X, Y, Z) with an angle parameter γ for representing C_n, and we assume that $C_n = A_n\gamma + B_n$. γ represents the angle between the OX axis of fish body and the line which connects the origin O with the pressure sensor on the surface of the robotic fish, as shown in Figure 4.2.

As a sequence, for the ALLS,

$$
\begin{aligned}
\Delta V_{dynamics}(t, \gamma) =& (A_1\gamma + B_1)U^2 + (A_2\gamma + B_2)\theta^2 \\
&+ (A_3\gamma + B_3)\omega^2 + (A_4\gamma + B_4)U\theta \\
&+ (A_5\gamma + B_5)U\omega + (A_6\gamma + B_6)\theta\omega \\
&+ (A_7\gamma + B_7)
\end{aligned}
\tag{4.10}
$$

Considering that it is difficult to determine A_n and $B_n (n = 1, 2, 3, 4, 5, 6, 7)$, massive kinematic experiments are conducted for determining A_n and B_n, and validating the accuracy of the PV model for the robotic fish in the following parts.

4.3.2 Pressure Variation Models for Multiple Motions of the Robotic Fish

In experiments, the PVs measured by ALLS are named $\Delta P_{k_{alls}} (k = r, t, g, s)$. They consists of two parts: hydrodynamic PVs $\Delta P_{k_{dynamics}}$ and static PVs $\Delta P_{k_{statics}}$. $\Delta P_{k_{alls}}$ is expressed as

$$
\Delta P_{k_{alls}} = \Delta P_{k_{dynamics}} + \Delta P_{k_{statics}}
\tag{4.11}
$$

$$\Delta P_{k_{statics}} = (A_{kh}\gamma + B_{kh})\Delta P_{ks} \tag{4.12}$$

where $k = r, t, g, s$. "r", "t", "g", and "s" represents rectilinear motion, turning motion, gliding motion, and spiral motion of the robotic fish, respectively. ΔP_{ks} represents the static PVs measured by $P_{statics}$ in corresponding motions. Because of the rhythmical oscillation of fish body, the static PVs at the positions where $P_{statics}$ and ALLS lie are different, so two coefficients A_{kh} and B_{kh} are used for compensating the difference between the static PVs of $P_{statics}$ and ALLS. In the following identification process of the model parameters, average U, average ω, and average θ in a certain period are used for identify the model parameters. For rectilinear motion and gliding motion of the robotic fish, considering that the sway angular velocity ω of fish body oscillates around zero, so it can be excluded in terms of the average statistical characteristics over a certain period. The PV model is expressed as

$$\begin{aligned}\Delta P_{\lambda_{dynamics}} =&(A_{\lambda 1}\gamma + B_{\lambda 1})U_\lambda^2 + (A_{\lambda 2}\gamma + B_{\lambda 2})\theta_\lambda^2 \\ &+ (A_{\lambda 3}\gamma + B_{\lambda 3})U_\lambda\theta_\lambda + (A_{\lambda 4}\gamma + B_{\lambda 4})\end{aligned} \tag{4.13}$$

where $\lambda = r, g$.

While for turning motion and spiral motion of the robotic fish, the PV model is expressed as

$$\begin{aligned}\Delta P_{\mu_{dynamics}} =&(A_{\mu 1}\gamma + B_{\mu 1})U_\mu^2 + (A_{\mu 2}\gamma + B_{\mu 2})\theta_\mu^2 \\ &+ (A_{\mu 3}\gamma + B_{\mu 3})\omega_\mu^2 + (A_{\mu 4}\gamma + B_{\mu 4})U_\mu\theta_\mu \\ &+ (A_{\mu 5}\gamma + B_{\mu 5})U_\mu\omega_\mu + (A_{\mu 6}\gamma + B_{\mu 6})\theta_\mu\omega_\mu \\ &+ (A_{\mu 7}\gamma + B_{\mu 7})\end{aligned} \tag{4.14}$$

where $\mu = t, s$.

In addition, for rectilinear motion and turning motion of the robotic fish, the depth variations are negligible, so the static PVs $\Delta P_{statics}$ are excluded. While $\Delta P_{statics}$ are considered for gliding motion and spiral motion of the robotic fish. As a result, the $\Delta P_{k_{alls}}$ are expressed as

$$\Delta P_{k_{alls}} = \begin{cases} \Delta P_{k_{dynamics}} & k = r, t \\ \Delta P_{k_{dynamics}} + \Delta P_{k_{statics}} & k = g, s \end{cases} \tag{4.15}$$

4.3.3 Identification Process of the Model Parameters

Take turning motion for example, as shown in Figures 4.3 and 4.4, ΔP, ω_t, and θ vary with the time. In turning motion, the robotic fish swam

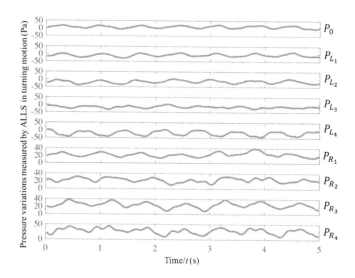

Figure 4.3 Real-time pressure variations $\Delta P_{t_{alls}}$ measured by ALLS in turning motion.

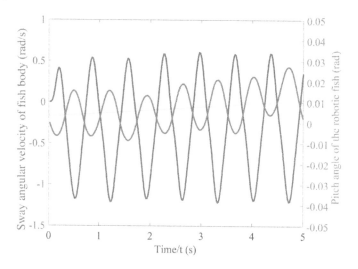

Figure 4.4 Real-time sway angular velocity ω_t and pitch angle θ of the robotic fish in turning motion.

anticlockwise. In this case, the PVs measured by $P_{R_m}(m = 1, 2, 3, 4)$ are positive, while the PVs measured by $P_{L_m}(m = 1, 2, 3, 4)$ oscillate around negative values. For identifying the PV model parameters, mean of ten-seconds $\Delta P_{k_{alls}}(k = r, t, g, s)$ measured by ALLS, mean of

TABLE 4.2 List of Parameters of the PV Models for Multiple Swimming Motions

Parameters	Value	Parameters	Value
A_{r1}	-2.8×10^2	A_{g1}	-8.1×10^2
A_{r2}	3.8×10^3	A_{g2}	81
A_{r3}	-5.2×10^3	A_{g3}	-2.9×10^3
A_{r4}	-1.8	A_{g4}	0.0024
A_{rh}	/	A_{gh}	26
B_{r1}	8.1×10^2	B_{g1}	9.8×10^2
B_{r2}	-6.7×10^3	B_{g2}	74
B_{r3}	9.9×10^3	B_{g3}	3.1×10^3
B_{r4}	2.0	B_{g4}	0.99
B_{rh}	/	B_{gh}	-23
A_{t1}	1.3×10^3	A_{s1}	1.2×10^3
A_{t2}	2.0×10^3	A_{s2}	-4.1×10^2
A_{t3}	1.2×10^2	A_{s3}	-9.6
A_{t4}	-1.0×10^4	A_{s4}	-4.2×10^2
A_{t5}	-9.1×10^2	A_{s5}	-6.3×10^2
A_{t6}	2.4×10^3	A_{s6}	-1.2×10^3
A_{t7}	-0.12	A_{s7}	0.0031
A_{th}	/	A_{sh}	25
B_{t1}	1.9×10^3	B_{s1}	-6.5×10^3
B_{t2}	0	B_{s2}	4.5×10^2
B_{t3}	-94	B_{s3}	89
B_{t4}	9.3×10^3	B_{s4}	2.5×10^3
B_{t5}	1.4×10^3	B_{s5}	2.7×10^3
B_{t6}	-9.8×10^2	B_{s6}	7.2×10^2
B_{t7}	-3.9	B_{s7}	0.99
B_{th}	/	B_{sh}	-55

ten-seconds $\Delta P_{k_{statics}}(k = r, t, g, s)$ measured by $P_{statics}$, mean of ten-seconds $U_k(k = r, t, g, s)$ measured by camera, mean of ten-seconds sway angular velocity $\omega_\mu(\mu = t, s)$ and mean of ten-seconds pitch angle $\theta_k(k = r, t, g, s)$ acquired by AHRS, and the angle parameter γ of each pressure sensor were imported into (4.13) and (4.14). The PV model parameters were identified using least square method based linear regression analysis, as shown in Table 4.2. Details about the identification were as follows.

Taking the hydrodynamic pressure model of spiral motion (4.16) for example,

$$
\begin{aligned}
\Delta P_{s_{dynamics}} =&(A_{s1}\gamma + B_{s1})U_s^2 + (A_{s2}\gamma + B_{s2})\theta_s^2 \\
&+ (A_{s3}\gamma + B_{s3})\omega_s^2 + (A_{s4}\gamma + B_{s4})U_s\theta_s \\
&+ (A_{s5}\gamma + B_{s5})U_s\omega_s + (A_{s6}\gamma + B_{s6})\theta_s\omega_s \\
&+ (A_{s7}\gamma + B_{s7})
\end{aligned}
\tag{4.16}
$$

(4.16) can be expressed as

$$
\begin{aligned}
\Delta P_{s_{dynamics}} =&A_{s1}\gamma U_s^2 + B_{s1}U_s^2 + A_{s2}\gamma\theta_s^2 + B_{s2}\theta_s^2 \\
&+ A_{s3}\gamma\omega_s^2 + B_{s3}\omega_s^2 + A_{s4}\gamma U_s\theta_s + B_{s4}U_s\theta_s \\
&+ A_{s5}\gamma U_s\omega_s + B_{s5}U_s\omega_s + A_{s6}\gamma\theta_s\omega_s + B_{s6}\theta_s\omega_s \\
&+ A_{s7}\gamma + B_{s7}
\end{aligned}
\tag{4.17}
$$

(4.18) can be expressed as

$$
\begin{aligned}
y =&b_1x_1 + b_2x_2 + b_3x_3 + b_4x_4 \\
&+ b_5x_5 + b_6x_6 + b_7x_7 + b_8x_8 \\
&+ b_9x_9 + b_{10}x_{10} + b_{11}x_{11} + b_{12}x_{12} \\
&+ b_{13}x_{13} + b_{14}
\end{aligned}
\tag{4.18}
$$

where

$$
\begin{cases}
b_1 = A_{s1} \\
b_2 = B_{s1} \\
b_3 = A_{s2} \\
b_4 = B_{s2} \\
b_5 = A_{s3} \\
b_6 = B_{s3} \\
b_7 = A_{s4} \\
b_8 = B_{s4} \\
b_9 = A_{s5} \\
b_{10} = B_{s5} \\
b_{11} = A_{s6} \\
b_{12} = B_{s6} \\
b_{13} = A_{s7} \\
b_{14} = B_{s7}
\end{cases}
\tag{4.19}
$$

$$\begin{cases}
x_1 = \gamma U_s^2 \\
x_2 = U_s^2 \\
x_3 = \gamma \theta_s^2 \\
x_4 = \theta_s^2 \\
x_5 = \gamma \omega_s^2 \\
x_6 = \omega_s^2 \\
x_7 = \gamma U_s \theta_s \\
x_8 = U_s \theta_s \\
x_9 = \gamma U_s \omega_s \\
x_{10} = U_s \omega_s \\
x_{11} = \gamma \theta_s \omega_s \\
x_{12} = \theta_s \omega_s \\
x_{13} = \gamma
\end{cases} \tag{4.20}$$

(4.18) is a multiple linear regression model $(y = f(x_1, x_2, \ldots, x_{13}))$. $y, x_1, x_2, \ldots, x_{13}$ can be acquired using the recording data in the experiments. Specifically, for identifying the PV model parameters, mean of ten-seconds $\Delta P_{k_{alls}}(k = r, t, g, s)$ measured by ALLS, mean of ten-seconds $\Delta P_{k_{statics}}(k = r, t, g, s)$ measured by $P_{statics}$, mean of ten-seconds $U_k(k = r, t, g, s)$ measured by camera, mean of ten-seconds sway angular velocity $\omega_\mu(\mu = t, s)$ and mean of ten-seconds pitch angle $\theta_k(k = r, t, g, s)$ acquired by AHRS, and the angle parameter γ of each pressure sensor were imported into (4.18). Then PV model parameters were identified using least square method based linear regression analysis. The coefficient of determination (R^2) and mean absolute error (MAE) between measured value and estimated value were used for evaluating the performances of the acquired PV models.

4.3.4 Pressure Variation Model-Based Motion Parameters Estimation

Basing on the acquired PV models, the velocity $U_k(k = r, t, g, s)$ of the robotic fish can be estimated by solving the PV models inversely using the ALLS-measured $\Delta P_{k_{alls}}(k = r, t, g, s)$ and the AHRS measured $\omega_\mu(\mu = t, s)$. That is to say, solving a linear system of the form $Ax = b$, where b are the ALLS and AHRS measurements (and their order-2 products), A are the coefficients determined by linear regression before, and x are the unknown velocities. Using the estimated $U_\mu(\mu = t, s)$, we can also estimate the turning radius R_T/spiral radius R_S of the robotic fish in turning motion and spiral motion. Specifically, R_T or R_S is the quotients of $U_\mu(\mu = t, s)$ and $\Omega_\mu(\mu = t, s)$, where $\Omega_\mu(\mu = t, s)$ is the

TABLE 4.3 Estimated Parameters Based on the Pressure Variations Model in the Four Motions

Experiments	Rectilinear motion	Turning motion	Gliding motion	Spiral motion
Estimated parameter	U_r	U_t, R_T	U_g	U_s, R_S

turning/spiral angular velocity of the robotic fish. For $\Omega_\mu(\mu = t, s)$ in a certain period (which specifically refer to one second in the real-time estimation), it can be estimated as the mean of the AHRS-measured sway angular velocity $\omega_\mu(\mu = t, s)$ in this period. Table 4.3 shows the estimated parameters based on the pressure variations model in the four motions. Basing on the estimated $U_k(k = r, t, g, s)$, R_T, R_S, and $\Omega_\mu(\mu = t, s)$, the trajectory of robotic fish can be calculated in the following part. For real-time estimation of the trajectory, real-time motion parameters including $U_k(k = r, t, g, s)$, R_T, R_S, and $\Omega_\mu(\mu = t, s)$ are recursively calculated by using one-second historical sensor data.

4.4 ARTIFICIAL LATERAL LINE-BASED TRAJECTORY ESTIMATION

For estimating the trajectory of the robotic fish, the coordinate of the robotic fish is calculated every second, using the motion parameters including ω_t and ω_s measured by the AHRS and $U_k(k = r, t, g, s)$, R_T, and R_S estimated by the proposed PV models. The trajectory estimation processes are shown in the attached video. In the following part, the trajectory of the robotic fish is described by its coordinate (x, y, z). $(x_{t_i}, y_{t_i}, z_{t_i})$ and $(x_{t_{i-1}}, y_{t_{i-1}}, z_{t_{i-1}})$ are coordinates of the robotic fish in $O_I X_I Y_I Z_I$ at time t_i and t_{i-1}, respectively. i indicates time series.

4.4.1 Trajectory Estimation of Rectilinear Motion

As shown in Figure 3.6(a), there exists rhythmical oscillation of fish head when the robotic fish is swimming. So the coordinate of the robotic fish is calculated with considering the orientation of the fish head. $\Delta\alpha$ is used to indicate orientation of the fish head relative to its initial orientation.

The coordinate of the robotic fish is finally expressed as

$$x_{t_i} = x_{t_{i-1}} + U_{r_{t_{i-1}t_i}} * \Delta t * cos(\Delta\alpha_{t_i})$$
$$y_{t_i} = y_{t_{i-1}} + U_{r_{t_{i-1}t_i}} * \Delta t * sin(\Delta\alpha_{t_i}) \qquad (4.21)$$
$$z_{t_i} = 0$$

where $\Delta t = t_i - t_{i-1} = 1s$.

4.4.2 Trajectory Estimation of Turning Motion

For a robotic fish in a steady state of turning motion, it comes back to initial position if the trajectory is a full circumference. Otherwise, for calculating the coordinate of a robotic fish whose trajectory is less than a full circumference, four cases are considered, as shown in Figure 4.5. The calculation of the turning trajectory is as follows.

Firstly, calculating the center of the turning trajectory. Here, we define the center of the turning trajectory in time interval (t_{i-1}, t_i) as $(xO_{t_{i-1}t_i}, yO_{t_{i-1}t_i})$, taking the form as

$$xO_{t_{i-1}t_i} = x_{t_{i-1}} - R_{T_{t_{i-1}t_i}} * sin(\Delta\alpha_{i-1})$$
$$yO_{t_{i-1}t_i} = y_{t_{i-1}} + R_{T_{t_{i-1}t_i}} * cos(\Delta\alpha_{i-1}) \qquad (4.22)$$
$$zO_{t_{i-1}t_i} = 0$$

where $R_{t_{i-1}t_i} = U_{t_{i-1}t_i}/\Omega_{t_{i-1}t_i}$. It is the turning radius of the robotic fish in time interval (t_{i-1}, t_i).

Secondly, calculating the coordinate of the robotic fish in time interval (t_{i-1}, t_i). Here, we define the coordinate of the robotic fish in (t_{i-1}, t_i) as $(x_{t_{i-1}t_ij}, y_{t_{i-1}t_ij})$, and $j = 1, 2, ..., 50$. $(x_{t_{i-1}t_ij}, y_{t_{i-1}t_ij})$ takes the form as

$$x_{t_{i-1}t_ij} = xO_{t_{i-1}t_i} + R_{T_{t_{i-1}t_i}} * cos\zeta_{t_{i-1}t_ij}$$
$$y_{t_{i-1}t_ij} = yO_{t_{i-1}t_i} + R_{T_{t_{i-1}t_i}} * sin\zeta_{t_{i-1}t_ij} \qquad (4.23)$$
$$z_{t_{i-1}t_ij} = 0$$

where $\zeta_{t_{i-1}t_ij}$ is the polar angle of the coordinate point $(x_{t_{i-1}t_ij}, y_{t_{i-1}t_ij})$. The range of $\zeta_{t_{i-1}t_ij}$ is defined as $(\zeta_{t_{i-1}t_il}, \zeta_{t_{i-1}t_ir})$, which is determined as shown in Table 4.4. $\zeta_{t_{i-1}t_il}$ and $\zeta_{t_{i-1}t_ir}$ are calculated by the yaw angle α_{t_i} at time t_i and α_{i-1} at time t_{i-1}.

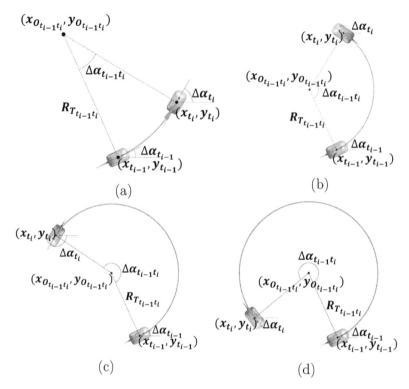

Figure 4.5 Four cases for calculating the coordinate of the robotic fish in turning motion. (a) Case 1: arc length between point (x_{t_i}, y_{t_i}) and point $(x_{t_{i-1}}, y_{t_{i-1}})$ is less than a quarter of circumference. (b) Case 2: arc length is more than a quarter of circumference and less than half a circumference. (c) Case 3: arc length is more than half a circumference and less than three-quarter circumference. (d) Case 4: arc length is more than three-quarter circumference and less than a full circumference.

4.4.3 Trajectory Estimation of Gliding Motion

As shown in Figure 3.6(b), the coordinate of the robotic fish is expressed as

$$
\begin{aligned}
x_{t_i} &= x_{t_{i-1}} + \sqrt{(U_{g_{t_{i-1}t_i}} * \Delta t)^2 - (\Delta h_{t_{i-1}t_i})^2} * cos(\Delta \alpha_{t_i}) \\
y_{t_i} &= y_{t_{i-1}} + \sqrt{(U_{g_{t_{i-1}t_i}} * \Delta t)^2 - (\Delta h_{t_{i-1}t_i})^2} * sin(\Delta \alpha_{t_i}) \qquad (4.24) \\
z_{t_i} &= z_{t_{i-1}} + \Delta h_{t_{i-1}t_i}
\end{aligned}
$$

TABLE 4.4 $(\zeta_{t_{i-1}t_i l}, \zeta_{t_{i-1}t_i r})$ for Time Interval (t_{i-1}, t_i) in Turning Motion and Spiral Motion

α_{t_i} \ $\alpha_{t_{i-1}}$	(0,90]	(90,180]	(−180,−90]	(−90,0]
(0,90]	$[\alpha_{t_{i-1}} - 90, \alpha_{t_i} - 90)$	$[\alpha_{t_{i-1}} - 90, \alpha_{t_i} + 270)$	$[\alpha_{t_{i-1}} + 270, \alpha_{t_i} + 270)$	$[\alpha_{t_{i-1}} - 90, \alpha_{t_i} - 90)$
(90,180]	$[\alpha_{t_{i-1}} - 90, \alpha_{t_i} - 90)$	$[\alpha_{t_{i-1}} - 90, \alpha_{t_i} - 90)$	$[\alpha_{t_{i-1}} + 270, \alpha_{t_i} + 270)$	$[\alpha_{t_{i-1}} - 90, \alpha_{t_i} - 90)$
(−180,−90]	$[\alpha_{t_{i-1}} - 90, \alpha_{t_i} + 270)$	$[\alpha_{t_{i-1}} - 90, \alpha_{t_i} + 270)$	$[\alpha_{t_{i-1}} + 270, \alpha_{t_i} + 270)$	$[\alpha_{t_{i-1}} - 90, \alpha_{t_i} + 270)$
(−90,0]	$[\alpha_{t_{i-1}} - 90, \alpha_{t_i} + 270)$	$[\alpha_{t_{i-1}} - 90, \alpha_{t_i} + 270)$	$[\alpha_{t_{i-1}} + 270, \alpha_{t_i} + 270)$	$[\alpha_{t_{i-1}} - 90, \alpha_{t_i} - 90)$

where $\Delta h_{t_{i-1}t_i}$ are the depth variations of the robotic fish in time interval (t_{i-1}, t_i), respectively.

4.4.4 Trajectory Estimation of Spiral Motion

For spiral motion, the x-coordinate and y-coordinate of the robotic fish are calculated in the same way as coordinates in turning motion. For z-coordinate of spiral motion, it is expressed as

$$z_{t_i} = z_{t_{i-1}} + \Delta h_{t_{i-1}t_i} \tag{4.25}$$

4.5 EXPERIMENTS

4.5.1 Rectilinear Motion

In rectilinear motion, the PVs measured by two corresponding pressure sensors at the left and right sides of the robotic fish were nearly equal. Thus, we used their average values and the PVs measured by P_0 to identify the parameters of the PV model and then inversely acquire the rectilinear velocity U_r basing on the identified PV model and the ALLS-measured PVs. Figures 4.6(a)–(e) and (f)–(j) show measured PVs acquired by ALLS and estimated PVs acquired by PV model for rectilinear motion, respectively. Figure 4.7 shows the measured and estimated velocity U_r acquired by reversely solving the PV model. The R^2 and MAE values for PV and U_r are calculated as shown in Table 4.5.

TABLE 4.5 R^2 and MAE Values for the Evaluated Parameters in Four Motions

Experiments	Evaluated parameters	MAE	R2
Rectilinear motion	PV	2.7 Pa	0.98
	U_r	0.00600 m/s	0.910
Turning motion	PV	1.9 Pa	0.97
	U_t	0.00520 m/s	0.874
	R_T	0.00810 m	0.979
Gliding motion	PV	17 Pa	0.99
Spiral motion	PV	20 Pa	0.99

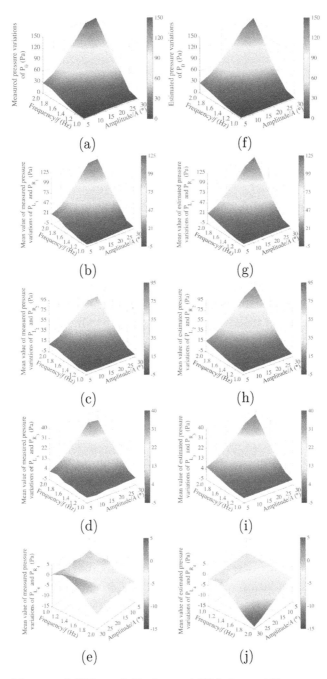

Figure 4.6 Measured PVs and Estimated PVs in rectilinear motion. (a), (b), (c), (d), and (e) Measured value. (f), (g), (h), (i), and (j) Estimated value.

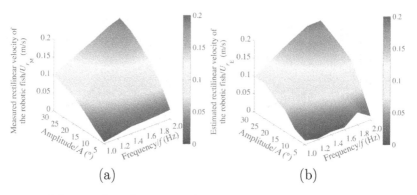

(a)   (b)

Figure 4.7 Measured and estimated rectilinear velocity of the robotic fish in rectilinear motion. (a) Measured value. (b) Estimated value.

In order to validate the acquired PV model, five experiments with given oscillating parameters of the tail which are different from those in Figure 4.8 have been conducted. Each experiment was conducted for five times, and the mean value of the PVs and mean value of the velocities for the five experiments were calculated. Figure 4.8 shows the measured and estimated PVs measured by ALLS in validation experiments. EP_0 and MP_0 respectively indicate the estimated and measured PVs measured by pressure sensor P_0. MEP_m and MMP_m $(m = 1, 2, 3, 4)$ respectively indicate mean value of estimated PVs and mean value of measured PVs of pressure sensor P_{R_m} $(m = 1, 2, 3, 4)$ and P_{L_m} $(m = 1, 2, 3, 4)$. The absolute error ΔV defined in (4.26) was used to estimate the accuracy of PVs. $V_{measured}$ and $V_{estimated}$ in (4.26) respectively indicate the measured and estimated value of PVs. The maximum and minimum absolute errors of PVs are 11 Pa and 0.89 Pa, respectively.

$$\Delta V = |V_{measured} - V_{estimated}| \tag{4.26}$$

Figure 4.9 shows the measured and estimated velocity U_r of the robotic fish in validation experiments. The percentage error δ defined in (4.27) was used to estimate the accuracy of the estimated U_r. $U_{measured}$ and $U_{estimated}$ in (4.27) respectively indicate the value of measured U_r and estimated U_r for each experimental parameter. The percentage error of U_r for each experiment is 10.2%, 6.73%, 0.450%, 0.650%, and 0.0900%, respectively.

$$\delta = \left| \frac{U_{measured} - U_{estimated}}{U_{measured}} \right| \times 100\% \tag{4.27}$$

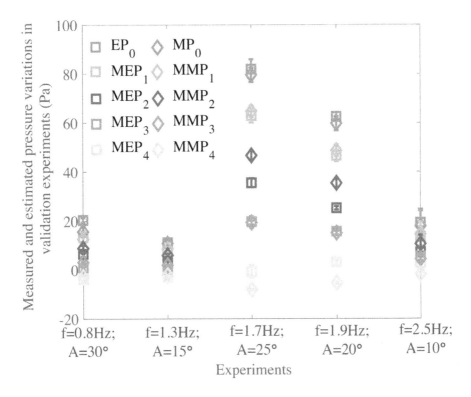

Figure 4.8 Measured and estimated PVs in validation experiments of rectilinear motion.

Figure 4.10 shows the measured and estimated trajectory of rectilinear motion in 10 seconds. The trends of the two trajectories match well. Figure 4.11 shows the position tracking error with the time. For the rectilinear motion, the error increase results from the approximate estimation of rectilinear velocity and accumulative error in calculation procedure. Nevertheless, the positions of the two end points have an error of only 0.166 m in 10 seconds, while the total path length of the robot is 1.76 m.

4.5.2 Turning Motion

In turning motion experiments, PVs measured by $P_{R_m}(m = 1, 2, 3, 4)$ were used for identifying the parameters of PV model and inversely acquiring the turning velocity U_t. Figure 4.12 shows measured PVs acquired by ALLS and estimated PVs acquired by PV model for turning motion. Figures 4.13 and 4.14 show the linear velocity U_t and turning

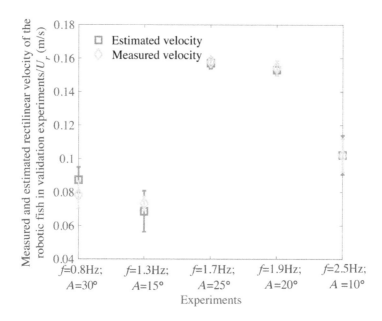

Figure 4.9 Measured and estimated rectilinear velocity of the robotic fish in verification experiments of rectilinear motion.

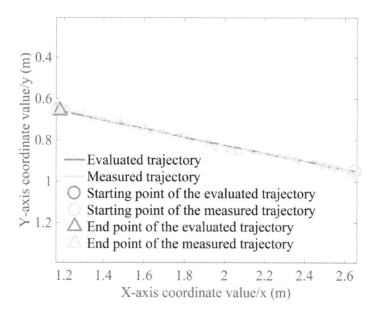

Figure 4.10 Rectilinear trajectory of the robotic fish in 10 seconds.

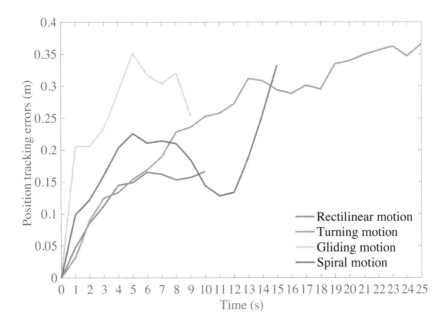

Figure 4.11   Position tracking errors of the robotic fish in four motions.

radius R_T of the robotic fish in turning motion, respectively. From R^2 and MAE values in Table 4.5, it can be concluded that the estimated PV, U_t, and R_T values match closely with their measured values.

The estimated turning velocity tracks the measured value well with an average absolute error of 0.0175 m/s and a maximum absolute error of 0.0774 m/s (Figure 4.15(a)). Comparing to the estimation result for velocity of a robotic fish in [116], the proposed model can also precisely estimate the low velocity. Figure 4.15(b) shows the real-time turning angular velocity Ω_t of the robotic fish. As we have mentioned before, it is acquired by recursively calculated mean of one-second historical sway angular velocities of AHRS. From $t = 6s$ on, the Ω_t oscillates around a certain value, which means the robotic fish is in steady state of turning motion. Figure 4.15(c) shows the real-time estimated turning radius of the robotic fish. At the starting period $(t = 0 - 2s)$, the turning radii are large enough that its motion can be regarded as rectilinear motion. From $t = 6s$ on, the R_{T_E} oscillate around a certain value. R_{T_E} is a quotient of estimated turning velocity in Figure 4.15(a) and turning angular velocity in Figure 4.15(b). As far as we know, few sensors can be used to measure real-time turning radius for underwater robots. In the experiments, we

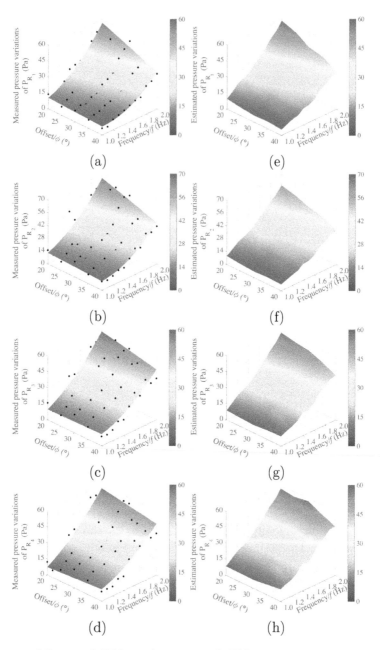

Figure 4.12 Measured PVs and estimated PVs in turning motion. (a), (b), (c), and (d) Measured value. (e), (f), (g), and (h) Estimated value. Black dots and the curved surfaces in (a), (c), (e), and (g) respectively represent actual values and least square method based fitted values of the measured PVs.

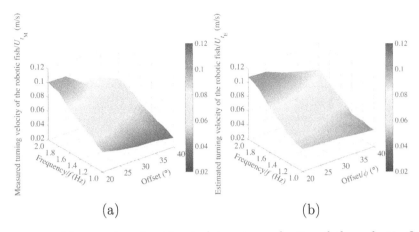

(a)　　　　　　　　　　(b)

Figure 4.13 Measured and estimated turning velocity of the robotic fish in turning motion. (a) Measured value. (b) Estimated value.

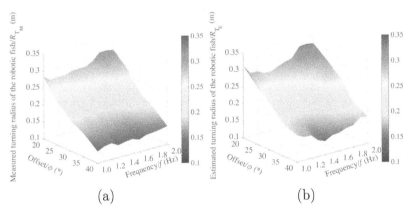

(a)　　　　　　　　　　(b)

Figure 4.14 Measured and estimated turning radius of the robotic fish in turning motion. (a) Measured value. (b) Estimated value.

cannot acquire real-time ground truth of turning radius for the robotic fish. Though it is possible to calculate the ground truth of turning radius using the camera-recored coordinates, the accuracy is too low. So Figure 18 (c) just shows the estimated values of turning radius. On the other hand, considering that the estimated linear velocity matches with the measured linear velocity with small errors, as shown in Figure 18 (a), it can be initially inferred that the real-time estimated turning radius has good agreement with the ground truth of turning radius. Besides, we have used the estimated turning radius to estimated the trajectory of the robotic fish, the small errors of estimated and measured trajectory

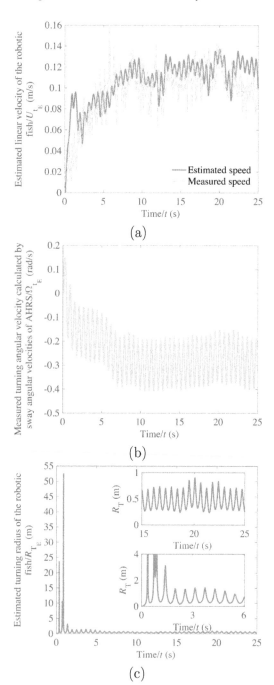

Figure 4.15 Real-time turning velocity U_t, turning angular veloctiy Ω_t, and turning radius R_T of the robotic fish in turning motion. (a) Real-time U_t. (b) Real-time Ω_t. (c) Real-time R_T.

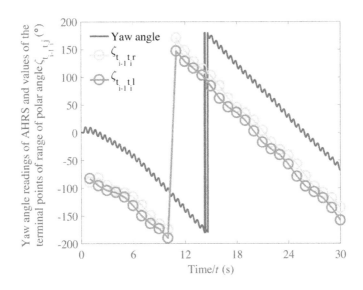

Figure 4.16 Yaw angle readings of AHRS and values of the terminal points of range of polar angle $\zeta_{t_{i-1}t_i}$.

further demonstrate the accuracy of the estimated turning radius. So this work also provides a promising way for obtaining the real-time turning radius for underwater robots, using the onboard ALLS and AHRS.

Figure 4.16 shows the real-time yaw angle of the robotic fish and terminal points of range of polar angle (Table 4.4) used for calculated the turning trajectory. Figure 4.17 shows the measured and estimated turning trajectory of the robotic fish in 25 seconds. As shown in Figure 4.11, comparing to the estimation of rectilinear trajectory, the estimation error of the turning trajectory is bigger, with an estimation error of 0.256 m after 10-seconds swimming. Considering that the original values of the PVs measured by the ALLS exist data mutation in large scopes of experimental parameter space, because of the pressure sensors' own defects, we have used the least square method based fitted value of the PVs for identifying the PV model. Such a process may result in the bigger estimation error. Besides, the position estimation error between the two end points is 0.362 m while the total path length of the robot is 2.51 m.

4.5.3 Gliding Motion

In gliding motion, we also use the average value of the PVs measured by two corresponding pressure sensors at the left and right sides and the

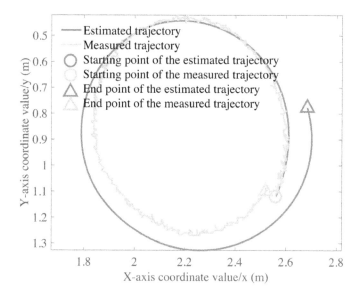

Figure 4.17 Turning trajectory of the robotic fish in 25 seconds.

PVs measured by P_0 to identify the parameters of the PV model and inversely acquire the gliding velocity U_g. Figure 4.18(a) shows the ALLS-measured PVs and the PV model-estimated PVs in gliding motion. As shown in Figure 4.18(b), errors between the estimated and measured values are much less than them. From R^2 and MAE values in Table 4.5, it can be concluded that the measured and estimated PV match closely. Four validation experiments are conducted for validating the PV model. The minimum and maximum percentage errors between the measured PVs and the estimated PVs in validation experiments are 0.67% and 0.056%, respectively. Figure 4.19 shows the measured and estimated gliding velocity U_g, the percentage errors between the measured values and corresponding estimated values of U_g vary from 6.46% to 14.9%. For the validation experiments, the percentage errors of the U_g are 2.08%, 1.91%, 2.76%, and 10.9% for $\Delta d = -1.3$ m, -1.0 m, -0.6 m, and -0.2 m, respectively. Figure 4.20 shows the trajectory of the robotic fish in gliding motion. The position tracking error of the gliding trajectory results from the estimated gliding velocity and the depth variations measured by the pressure sensor $P_{statics}$. As shown in Figure 4.11, the error between the ending points of the measured trajectory and the estimated trajectory is 0.253 m, while the total path length of the robot is 2.09 m.

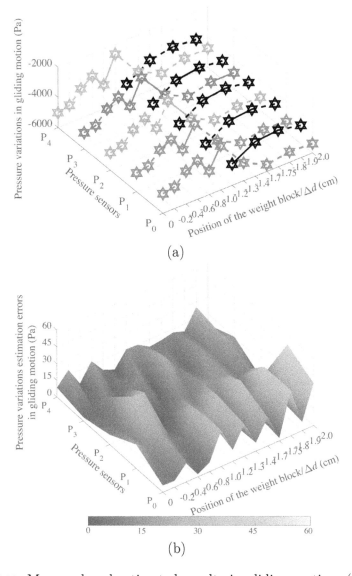

(a)

(b)

Figure 4.18 Measured and estimated results in gliding motion. (a) PVs in gliding motion. The values corresponds to P_m $(m = 1, 2, 3, 4)$ indicates the mean value of P_{L_m} and P_{R_m}. The black markers mean the PVs in validation experiments, while the markers in other colors mean the measured PVs used for identification of the PV model and their corresponding estimated values. The markers '$\triangle$' mean the estimated values of the PVs while the markers '$\triangledown$' mean the measured values. (b) PVs estimation errors.

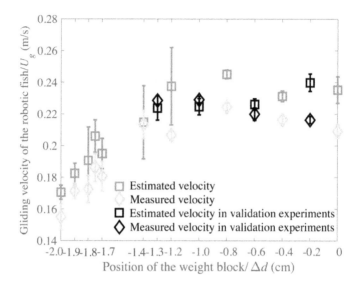

Figure 4.19 Measured and estimated gliding velocity of the robotic fish in gliding motion.

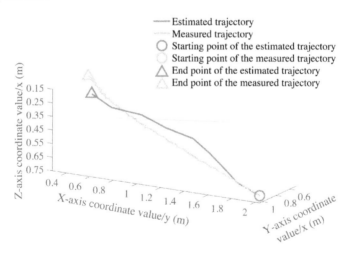

Figure 4.20 Gliding trajectory of the robotic fish in 9 seconds.

4.5.4 Spiral Motion

In spiral motion experiments, the robotic fish also turned anticlockwise, so PVs measured by $P_{R_m}(m = 1, 2, 3, 4)$ were used for identifying the parameters of PV model and inversely acquiring the spiral velocity U_s, as the case for turning motion. Three validation experiments are conducted for validating the accuracy of the PV model and its efficiency in estimating the spiral velocity U_s. Figure 4.21 shows the measured PVs and estimated PVs in spiral motion. From R^2 and MAE values in Table 4.5, it can be concluded that the the measured and estimated PV match closely. Besides, the minimum and maximum percentage errors between the measured values and the estimated values vary from 0.099% to 1.7% in validation experiments. Figure 4.22 shows the measured and estimated spiral velocity U_s. The percentage errors between the measured values and corresponding estimated values of U_s are 10.7%, 6.25%, 13.6%, 7.14%, and 1.24% for $\Delta d = 0$, -0.4 m, -0.6 m, -1.2 m, and -1.6 m, respectively. For the validation experiments, the percentage errors of the U_s are 7.28%, 13.3%, and 3.44% for $\Delta d = -0.2$ m, -1.0 m, and -1.4 m, respectively. Figure 4.23 shows the measured and estimated spiral radius R_S, the percentage errors between the estimated and measured R_S vary from 4.33% to 24.6%. Figure 4.24 shows the trajectory of the robotic fish in spiral motion. Figure 4.11 shows the real-time position tracking error of the robotic fish in spiral motion. Taking account of the trajectory estimation method in section 4.4.2, the position tracking error of the spiral trajectory results from the estimated spiral radius, the yaw angle variations measured by the AHRS, and the depth variations measured by the pressure sensor $P_{statics}$. Overall, the estimated trajectory reflects the evolution law of the measured trajectory well with an error of 0.333 m between the ending points of the measured trajectory and the estimated trajectory, while the total path length of the robot is 2.08 m.

4.5.5 Discussion

The experiments were all conducted under laboratory conditions, without considering the practicability of pressure sensor arrays based ALLS in open water such as river or lake. The key of application of pressure sensor arrays based ALLS in the wild is calibration of the pressure sensors. In the lab, the PVs measured by all the pressure sensors were calibrated relative to the pressure recordings in still water. However, calibration of the pressure sensors in the wild will be difficult and hard to setup, taking account of the complex underwater environment in the field (including uncertain turbulent flow, water temperature, atmospheric pressure, etc.). Nevertheless, some literatures have introduced the calibration of the pressure sensors [13, 137, 138]. In [13], the authors have introduced a generic calibration procedure for a pressure sensor

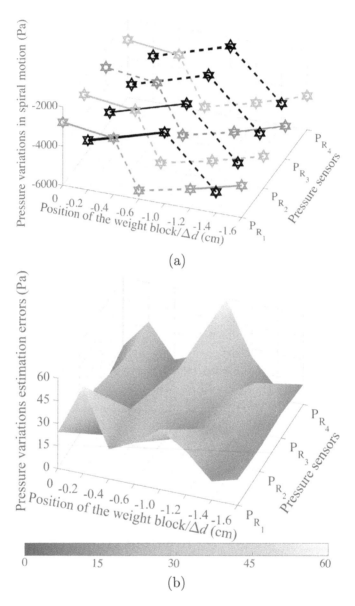

(a)

(b)

Figure 4.21 Measured and estimated results in spiral motion. (a) PVs in spiral motion. The black markers mean the PVs in validation experiments, while the markers in other colors mean the measured PVs used for identification of the PV model and their corresponding estimated values. The markers '△' mean the estimated values of the PVs while the markers '▽' mean the measured values. (b) PVs estimation errors.

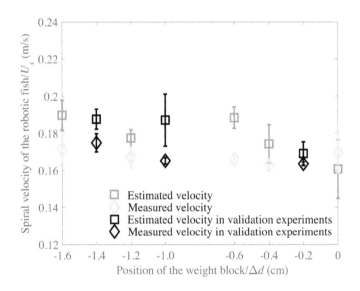

Figure 4.22 Measured and estimated spiral velocity of the robotic fish in spiral motion.

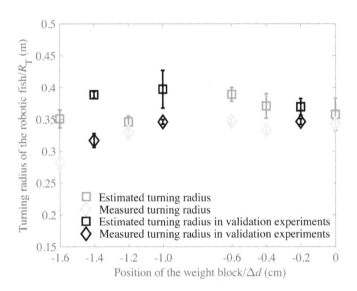

Figure 4.23 Measured and estimated spiral radius of the robotic fish in spiral motion.

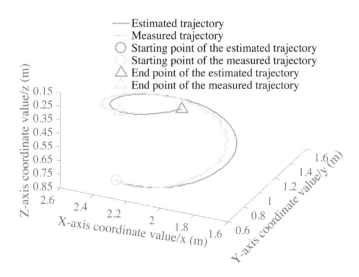

Figure 4.24 Spiral trajectory of the robotic fish in 15 seconds.

arrays based artificial lateral line system. The procedure includes calibration for temperature, atmospheric pressure, and water depth, and normalization for sensor bias. Specifically, the authors have established some calibration formulas to calibrated the pressure signal. Besides, all the pressure readings have been normalized by a specific sensor in order to make sure the identical performance of all the pressure sensors. In [137], the authors have presented a neural network based temperature compensation system for pressure sensor. It has been demonstrated that the system can efficiently remove the temperature drift and improve the accuracy of pressure measurement. In [138], the authors have established a Support Vector Machine (SVM) based calibration model to remove temperature and voltage fluctuation influence on the pressure sensor. The above-mentioned research has provided guidance for calibration of the pressure sensors in the wild. We believe that it is promising to realize online state estimation using the pressure sensors based ALLS in the wild.

4.6 CONCLUSIONS AND FUTURE WORK

Theoretical models which reflect PVs distributed on the surface of a freely-swimming robotic fish in rectilinear motion, turning motion, gliding motion, and spiral motion were respectively proposed. Basing on the proposed PV models, the motion parameters including the movement velocities and the turning/spiral radius of fish body were efficiently estimated using the ALLS-measured PVs. Using the estimated motionparameters, the trajectories of the robotic fish in three-dimensional space were estimated with small estimation error. This work demonstrates the effectiveness of artificial lateral line in state estimation of underwater robots and vehicles, indicating that ALLS is promising to be an essential complement to the usual sensor system which is equipped with underwater robots and vehicles. In the future work, we will focus on the calibration of the pressure sensor arrays based ALLS in the wild, and attempt to improve the practicability of the ALLS in natural conditions. We will design trajectory control loop, and input the ALLS-estimated trajectory as a feedback term, for realizing flow-aided online trajectory control of the robotic fish in the wild.

CHAPTER 5

Artificial Lateral Line-Based Local Sensing between Two Adjacent Boxfish-Like Robots

L ATERAL LINE SYSTEM (LLS) is a mechanoreceptive organ system with which fish and aquatic amphibians can effectively sense the surrounding flow field. The reverse Kármán vortex street (KVS), known as a typical thrust-producing wake, is commonly observed in fish-like locomotion and is known to be generated by fish tail. The vortex street generally reflects the motion information of fish. Fish can use LLS to detect such vortex streets generated by its neighboring fish, thus sensing its own states and the states of its neighbors in fish school. Inspired by such a typical biological phenomenon, we design a robotic fish with an onboard artificial lateral line system (ALLS) composed of pressure sensor arrays and use it to detect the reverse KVS-like vortex wake generated by its adjacent robotic fish. Specifically, the vortex wake results in the hydro-dynamic pressure variation (HPV) existed in flow field. Through measuring the HPV using ALLS and extracting meaningful information from the pressure sensor readings, the oscillating frequency/amplitude/offset of the adjacent robotic fish, the relative vertical distance and the relative yaw/pitch/roll angle between the robotic fish and its neighbor are

DOI: 10.1201/b23027-5

sensed efficiently. This work investigates the hydrodynamic characteristics of the reverse KVS-like vortex wake using ALLS. Besides, this work demonstrates the effectiveness and practicability of artificial lateral line in local sensing for adjacent underwater robots, indicating that it is promising to improve the close-range interaction and the cooperation for a group of underwater vehicles with the application of ALLS in the near future.

5.1 INTRODUCTION

The existing works have mainly studied the application of ALLS on one individual robot, rarely investigating an underwater robot group composed of two or more individuals with ALLS. On the other hand, the flow stimuli used in laboratory researches on ALLS typically include KVS behind a regular geometry model such as cylinder or cuboid [28, 139], wave generated by a vibrating sphere [7, 80, 98, 107, 140–142] and uniform flow generated by well controlled laboratory conditions [12, 19, 84]. Though these stimuli could well emulate the rhythmic movement of fins and fish body, they have the limitations to emulate the spatial and temporal hydrodynamic characteristics which natural flow stimuli have. Little attention has been drawn on natural flow stimuli which fish's LLS is subject to. The reverse KVS, known as a typical thrust-producing wake, is one of the above-mentioned natural stimuli. It is commonly observed in fish-like locomotion and is known to be generated by fish tail [143]. Besides, the vortex street generally reflects the motion information of the fish. Most species of fish are able to use their LLSs to detect such vortex streets generated by their neighboring fish, thus sensing the states of their neighbors [144]. Recently, a variety of bio-inspired robotic fish have attracted increasing attention largely because they perfectly imitate the hydrodynamic characteristics and swimming properties of their counterparts in nature. In particular, a swimming robotic fish is able to generate vortex wake which is similar to the reverse KVS produced by real fish [145]. Consequently, it can be used as a scientific and appropriate tool to generate the above-mentioned more natural flow stimulus in the study of lateral line-based sensing.

On the basis of the above-mentioned analyzes and inspired by the biological phenomenon of lateral line-based sensing existed in real fish, we mainly focus on using a bio-inspired robotic fish with an onboard ALLS composed of pressure sensors array to investigate the hydrodynamic characteristics of the reverse KVS-like vortex wake generated by its adjacent robotic fish. Specifically, computation fluid dynamics (CFD) simulation is utilized to preliminarily explore the distribution characteristics of the vortex wake. Then multiple experiments are conducted inside a low-turbulence flume in order to investigate the hydrodynamic pressure variations (HPVs) caused by the vortex wake. And by extracting meaningful information from the pressure sensor readings, the HPVs which reflect the relative states between the robotic fish and its neighbor are investigated efficiently. In our previous works, we have utilized a robotic fish with an onboard ALLS to sense the relative lateral distance and the relative longitudinal distance to its neighbor [24, 146]. We further apply ALLS to sense the relative vertical distance between the robotic fish and its neighbor. As shown in Figure 5.1, the relative lateral distance and the relative vertical distance between two adjacent robotic fish refer to the distance between two centers of mass along the X-axis and Z-axis, respectively. Besides, the relative longitudinal distance refers to the distance between the end of the caudal fin of the upstream robotic fish and the nose of the downstream robotic fish. Then we investigate using a robotic fish with an onboard ALLS for sensing the oscillating amplitude, frequency and offset of its neighbor. Finally, we explore using a robotic fish with an onboard ALLS to sense the relative yaw, pitch and roll angle to its neighbor. The yawing motion, pitching motion, and rolling motion of the robotic fish are also defined in Figure 5.1.

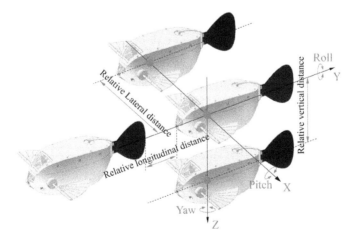

Figure 5.1 Relative positions between two adjacent robotic fish and the definition of the attitude motions. The red points indicate the centers of mass of the robotic fish.

5.2 MATERIALS AND METHODS

5.2.1 The Experimental Description

5.2.1.1 Individual Differences among Pressure Sensors

In order to ensure that the pressure variations measured by the nine pressure sensors are comparable, we have detailedly investigated the individual differences among the pressure sensors before conducting experiments. Specifically, we firstly gave a certain depth variation which corresponds to a certain pressure variation to each pressure sensor in still water. Then we compared the pressure variations measured by the pressure sensors. The concrete process of measuring the above-mentioned pressure variations are detailedly described as follows. Before conducting experiments, individual differences among the nine pressure sensors were analyzed in detail. Specifically, the individual differences were investigated by analyzing the pressure variations measured by the nine pressure sensors with respect to the depth variations of the robotic fish in still water. The depth variations of the robotic fish were adjusted by a position control setup which was built above a flume, as shown in Figure 5.2. The setup was able to separately control the positions of the robotic fish along the X-axis, Y-axis, and Z-axis with a position accuracy of 1 mm. In the experiment, the initial position of the robotic fish was fixed centrally within the flume, and the upper surface of the robotic fish just coincided with the surface of the water. Through controlling the position of the robotic fish along the Z-axis, the depth variation of the robotic fish changed from 1 mm to 20 mm with a interval of 1 mm. When the robotic fish was at the initial position, the mean value of the pressure data recorded and the depth were denoted as V_1 and D_1. When the robotic fish was at a given depth D_2, the mean value of the pressure data recorded was denoted as V_2. The pressure difference between V_1 and V_2 ($\Delta V = V_2 - V_1$) reflected the pressure variation caused by the depth variation ($\Delta D = D_2 - D_1$) of the robotic fish.

As shown in Figure 5.2, the nine pressure sensors measured nearly the same pressure variations when they were given the same depth variations in still water. It meant that they had no individual differences in measuring pressure variations. We have also found that slight drifts occurred when the duration of recording was longer than three minutes. This was probably due to the electronics or environmental changes, such as temperature and atmospheric pressure. Consequently, in order to minimize this effect, the duration of each recording was kept short (less than

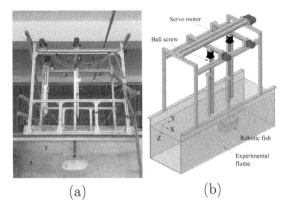

(a)   (b)

Figure 5.2 Position control setup with the robotic fish. (a) Isometric view. (b) Front view.

two minutes) in the following experiments. If the drift between the start and the end of the recording was too big, we repeated the recording.

5.2.1.2 The Experimental Platform

All the following experiments were conducted inside a low-turbulence flume in the State Key Laboratory for Turbulence and Complex Systems, Peking University, as shown in Figure 5.3. The experiments were conducted at the main test section whose size (Length × Width × Height) is about 600 cm × 40 cm × 50 cm. Besides, an axial-flow pump is used to create the circulation of the flow inside the flume. By adjusting the rotational speed, the flume provides an adjustable laminar flow. Before conducting experiments, the laminarity and velocity of the flow were checked by the digital particle image velocimetry (DPIV) technology at different rotational speeds of the axial-flow pump. In the experiments, the rotational speed was fixed at 200 revolutions per minute (rpm), and the corresponding flow velocity was fixed at about 17.5 centimeters per second (cm/s).

5.2.1.3 The Experimental Principle

It is well known that an oscillating fin results in generating forward propulsion and meanwhile backward reaction accompanying with the vortex wake at the downstream region. For two adjacent caudal fin-actuated robotic fish which are in leader-follower formation, the force between both of them mainly refers to the caudal fin-generating

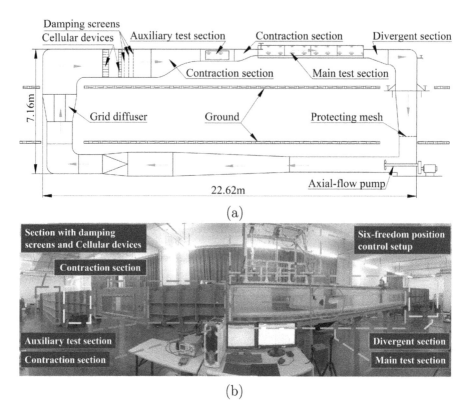

Figure 5.3 An introduction of the experimental flume. (a) Components of the experimental flume. (b) Overall view of the experimental flume above the floor. The red arrow indicates the direction of the flow. The robotic fish is highlighted by the circle.

backward reaction from the leader. From the point of force, the relative states between such two adjacent bio-inspired robotic fish can be approximately simplified to the relative states between a bio-inspired robotic fish and its adjacent individual oscillating caudal fin, as shown in Figure 5.4. Three types of experiments were conducted for sensing the above-mentioned relative states. Specifically, the experiments were classified as follows: 1) sensing the relative positions which specifically refers to sensing the relative vertical distance between the robotic fish and its adjacent individual oscillating caudal fin, 2) sensing the relative motions which specifically refers to sensing the oscillating states (amplitude, frequency, and offset) of the individual caudal fin, and 3) sensing the relative attitudes which specifically refers to sensing the relative yaw

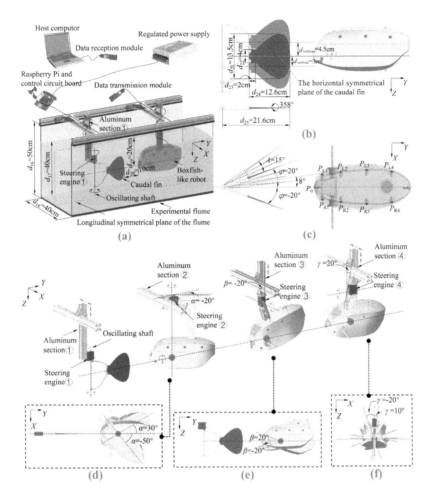

Figure 5.4 The diagrammatic sketch of the experiments. (a) The overall experimental apparatuses. The direction of the flow is along the Y-axis. (b) Sensing the relative vertical distance $d_{vertical}$. (c) Sensing the oscillating states of the adjacent caudal fin. A, f, and φ indicate the oscillating amplitude, frequency, and offset, respectively. (d) Sensing the relative yaw angle. (e) Sensing the relative pitch angle. (f) Sensing the relative roll angle. The red points in (d), (e), and (f) indicate the centers of mass of the robotic fish. α, β, and γ indicate the relative yaw angle, relative pitch angle, and relative roll angle which were adjusted by the output angles of the steering engines ②, ③, and ④.

angle, pitch angle and roll angle between the robotic fish and its adjacent individual oscillating caudal fin.

5.2.1.4 The Experimental Parameters

As shown in Figure 5.4(a), in all the experiments, the oscillating shaft of the oscillating caudal fin and the center of mass of the robotic fish were both fixed on the longitudinal symmetrical plane of the flume. Besides, when the yaw angle, pitch angle, and roll angle of the robotic fish were all 0°, the relative longitudinal distance between P_0 and the end of the caudal fin was 10 cm. In addition, the relative vertical distance between P_0 and the surface of the water was 20 cm. In the experiments of sensing the oscillating states of adjacent caudal fin and sensing the relative vertical distance between a robotic fish and adjacent caudal fin, the yaw angle, pitch angle and roll angle of the robotic fish were all fixed at 0°. And we defined the distance between P_0 and the horizontal symmetrical plane of the upstream caudal fin as the "relative vertical distance between a robotic fish and adjacent caudal fin". When the pressure sensor P_0 was on the horizontal symmetrical plane of the upstream caudal fin, the relative vertical distance was defined as zero. When P_0 was above the horizontal symmetrical plane, the relative vertical distance was defined as a positive value. Otherwise, it was defined as a negative value. By adjusting the aluminum section ① along the Z-axis in Figure 5.4(a), the relative vertical distance changed from −4.5 cm to 4.5 cm with an interval of 1.5 cm, as shown in Figure 5.4(b). Except for the experiment of sensing the relative vertical distance, the relative distance was fixed at zero in all of the other experiments. As shown in Figure 5.4(c), when the initial direction of the caudal fin was to the left of the downstream robotic fish, the oscillating offset was defined as a positive value. Otherwise, it was defined as a negative value. In the experiments of sensing the relative yaw, pitch and roll angles, the robotic fish was located with various relative yaw, pitch, and roll angles to its adjacent oscillating caudal fin, as shown in Figure 5.4(d–f). Besides, by respectively moving the aluminum sections ②, ③, and ④ along the X-axis, the Y-axis, and the Z-axis, the relative lateral distance, longitudinal distance and vertical distance between the center of mass of the robotic fish and the end of the upstream caudal fin remained unchanged. Other specific experimental parameters are listed in Table 4.1.

5.2.1.5 The Experimental Procedures

The experimental procedures were described as follows. First, locate an individual steering engine-driven caudal fin and a robotic fish with given relative positions and relative attitudes inside the flume. Second, record the pressure data in steady flow for thirty seconds, the mean value of the pressure data recorded was denoted as V_1. Third, stop the recording and then start the steering engine connected with the individual caudal fin with given control parameters for realizing given oscillating amplitude, frequency and offset, thus producing the vortex wake at the downstream region and meanwhile resulting in HPVs in the flow field. Then, wait for ten seconds until the vortex wake became stable. Then, record the pressure data for two minutes and the mean value of the pressure data recorded was denoted as V_2. Finally, stop the recording and then stop the steering engine. The difference between V_1 and V_2 ($\Delta V = V_2 - V_1$) reflected the HPV caused by the oscillating caudal fin. For each vertical distance, each oscillating amplitude/frequency/offset, and each relative yaw/pitch/roll angle, the above corresponding recordings were repeated for five times. Then the curves of the HPVs with respect to the relative vertical distance, the oscillating frequency/amplitude/offset of adjacent caudal fin, and the relative yaw/pitch/roll angle were obtained.

5.2.2 Computation Fluid Dynamics Simulation

As described previously, the vortex wake results in HPVs in flow field. In order to provide an important priori knowledge for the HPVs in the flow field and provide guidance for analyzing the HPVs, we have preliminarily explored the distribution characteristics of the vortex wake by conducting computation fluid dynamics (CFD) simulation using Fluent software. In the CFD simulation, we have selected the related actuation parameters, position parameters, and attitude parameters taking account of the corresponding parameters in experiments. Thus, the oscillating amplitude, frequency, and offset of the caudal fin are fixed at 15°, 1 Hz, and 0°, respectively. Besides, the relative yaw angle, pitch angle, and roll angle between the oscillating caudal fin and the robotic fish are all 0°. Additionally, the relative lateral distance, the relative vertical distance, and the relative longitudinal distance are 0 cm, 0 cm, and 10 cm, respectively. In addition, the flow velocity was fixed at 17.5 cm/s.

As shown in Figure 5.5, a vortex starts to form at the moment when the oscillating direction of the caudal fin reverses. Then it gradually develops and finally sheds when the oscillating caudal fin reaches the limit

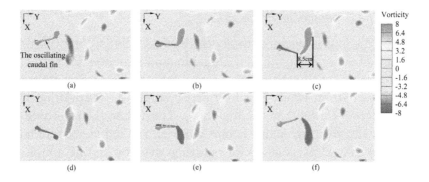

Figure 5.5 Vorticity maps in one oscillating period. (a) The caudal fin reverses the oscillating direction and a new vortex starts to form. (b) The caudal fin continues oscillating and the vortex continues developing. (c) The caudal fin reaches the limit position at the other side. The vortex finishes developing and finally sheds. (d), (e), and (f) repeat the processes of (a), (b), and (c).

position at the other side. In one oscillating period, two vortices shed from the end of the oscillating caudal fin with opposite rotation directions and the two shedding vortices are staggered up and down at the downstream region. Moreover, the oscillating amplitude determines the relative lateral distance between the two shedding vortices. As shown in Figure 5.6, the vorticity has an uneven distribution in three-dimensional space, just as described in [147]. The strongest vorticity appeared on the horizontal symmetrical plane of the caudal fin. Then it gradually decreases from the middle to both ends of the caudal fin along the Z-axis. Thus, it can be inferred that the biggest absolute value of the HPV exists on the horizontal symmetrical plane of the caudal fin. Moreover, with the increasing vertical distance to the horizontal symmetrical plane of the caudal fin, the absolute value of the HPV gradually decreases. Figure 5.7 shows an instantaneous vortex core region behind an individual oscillating caudal fin. As shown in Figure 5.7(c), the vortex structure is symmetrical with respect to the horizontal symmetrical plane of the caudal fin. Thus, it can be inferred that such a structure results in a symmetrical distribution of HPVs with respect to the horizontal symmetrical plane of the caudal fin. Besides, with the increasing longitudinal distance to the caudal fin, the shedding vortices gradually diffuse. As a result, the absolute values of the HPVs gradually decreases from the head to the end of the robotic fish, as shown in Figure 5.8. In addition, as shown

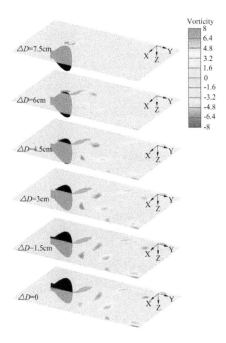

Figure 5.6 Instantaneous vorticity maps on different horizontal planes. ΔD indicates the distance (along the Z-axis) to the horizontal symmetrical plane of the caudal fin.

in Figure 5.8, the right and left sides of the robotic fish are subject to the same alternative positive and negative HPVs in one oscillating period. As a result, the HPVs measured by two corresponding pressure sensors at two sides were nearly equal in Experiment 1, 2, 3, and 6 (See Figures 5.9, 5.10, 5.11, and 5.14). Thus we mainly focused on analyzing the average value of the HPVs measured by two corresponding pressure sensors at two sides for these experiments. In the experiment of sensing the oscillating offset, the increasing offsets also resulted in the deviation of the shedding vortices' propagating direction from the direction of the flow. In this case, the vortex wake generated by the oscillating caudal fin may be dispersed by the flow before propagating downstream. Consequently, the hydrodynamic pressure-bearing area were concentrated on the head of the body of the robotic fish. Thus we have mainly analyzed the HPVs measured by P_0, P_{L_1}, and P_{R_1} in the article in order to acquire significant qualitative and quantitative relationships between the HPVs and the oscillating offsets.

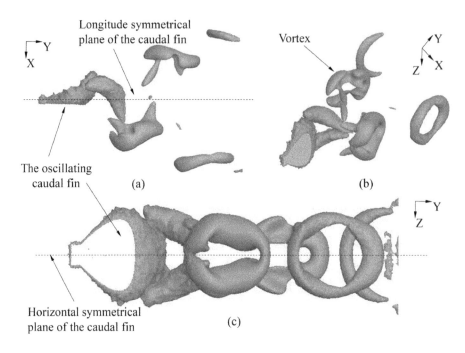

Figure 5.7 Instantaneous vortex core region behind an individual oscillating caudal fin. (a) Top view. (b) Isometric view. (c) Side view.

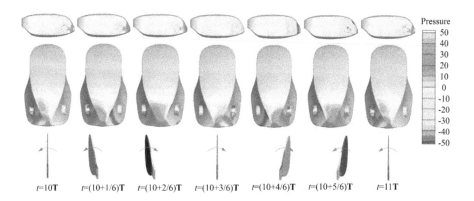

Figure 5.8 Hydrodynamic pressure variation caused by an individual oscillating caudal fin in computation fluid dynamics simulation based on Fluent software. The red arrows indicate the oscillating direction of the caudal fin. t indicates the time, and **T** indicates the oscillating period of the caudal fin.

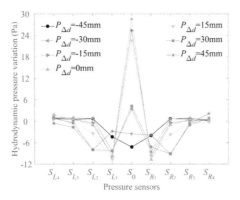

Figure 5.9 Hydrodynamic pressure variation measured by each pressure sensor in *Experiment 1*: Sensing the relative vertical distance between a robotic fish and its adjacent oscillating caudal fin.

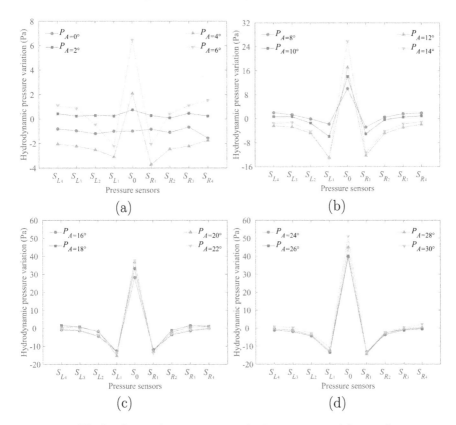

Figure 5.10 Hydrodynamic pressure variation measured by each pressure sensor in *Experiment 2*: Sensing the oscillating amplitude of adjacent oscillating caudal fin.

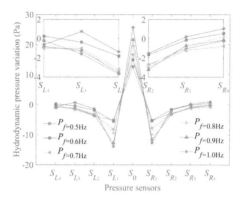

Figure 5.11 Hydrodynamic pressure variation measured by each pressure sensor in *Experiment 3*: Sensing the oscillating frequency of adjacent oscillating caudal fin.

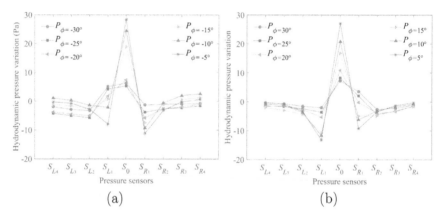

Figure 5.12 Hydrodynamic pressure variation measured by each pressure sensor in *Experiment 4*: Sensing the oscillating offset of adjacent oscillating caudal fin.

5.3 EXPERIMENTS

5.3.1 Experiment 1: Sensing the Relative Vertical Distance between a Robotic Fish and Its Adjacent Oscillating Caudal Fin

According to Bernoulli's principle, an increasing flow velocity leads to a decreasing hydrodynamic pressure [148]. Considering that the shedding vortices resulted in a increasing local flow velocity in the flow field, therefore, the shedding vortices resulted in a negative HPV. We define such an effect as the "negative effect" on the HPV. Besides, the shedding

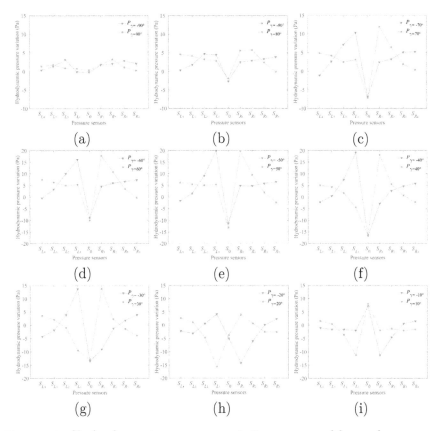

Figure 5.13 Hydrodynamic pressure variation measured by each pressure sensor in *Experiment 5*: Sensing the relative yaw angle between a robotic fish and its adjacent oscillating caudal fin.

vortices also resulted in backward reaction which is accompanied with the jet-flow in the vortex wake. It resulted in a positive HPV, and we define such an effect as the "positive effect" on the HPV. Moreover, the scope of the backward reaction mainly existed right behind the caudal fin. As shown in Figure 5.16, because of the negative effect caused by the vortices propagated along two sides of the robotic fish, the HPVs measured by the pressure sensors at the two sides maintained negative. By contrast, the HPV measured by P_0 maintained positive because of the positive effect caused by the backward reaction. For P_0 and P_1 (P_{L_1} and P_{R_1}), the HPV curves were approximately symmetrical about the axis $d_{vertical} = 0$. Because of the uneven distribution of vorticity behind the caudal fin as previously described in the section of computational fluid

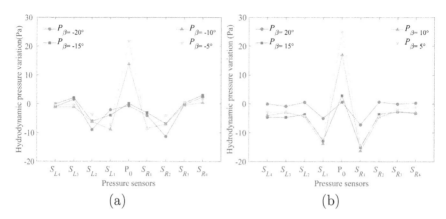

(a) (b)

Figure 5.14 Hydrodynamic pressure variation measured by each pressure sensor in *Experiment 6*: Sensing the relative pitch angle between a robotic fish and its adjacent oscillating caudal fin.

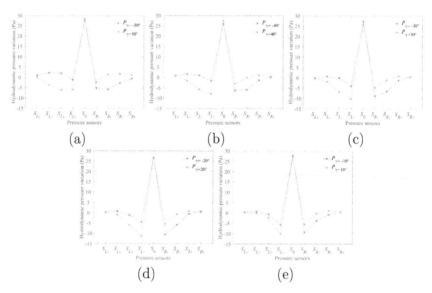

(a) (b) (c)

(d) (e)

Figure 5.15 Hydrodynamic pressure variation measured by each pressure sensor in *Experiment 7*: Sensing the relative roll angle between a robotic fish and its adjacent oscillating caudal fin.

dynamics simulation, the absolute values of the HPVs measured by P_0 and P_1 (P_{L_1} and P_{R_1}) reached the maximum values when $d_{vertical} = 0$. Moreover, with the increasing absolute value of relative vertical distance, the absolute values of the HPVs gradually decreased. For P_2 (P_{L_2} and

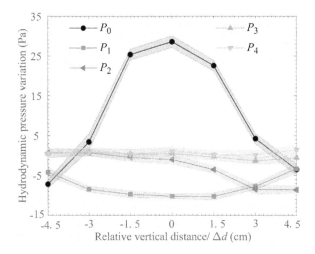

Figure 5.16 Hydrodynamic pressure variation with respect to the relative vertical distance. The value of P_n equals to the average value of P_{L_n} and P_{R_n} (n=1, 2, 3, and 4).

P_{R_2}), with the increasing relative vertical distance, they gradually approached to the horizontal symmetrical plane of the upstream caudal fin where the strongest vorticity existed. Consequently, the absolute values of the HPVs measured by P_2 (P_{L_2} and P_{R_2}) gradually increased. It was perhaps because the shedding vortices gradually diffused before approaching P_3 (P_{L_3} and P_{R_3}) and P_4 (P_{L_4} and P_{R_4}) that the HPVs measured by these pressure sensors were close to zero.

5.3.2 Experiment 2: Sensing the Oscillating Amplitude of Adjacent Oscillating Caudal Fin

As shown in Figure 5.17, the HPV measured by P_0 increased with the increasing oscillating amplitude. Besides, the HPVs measured by P_1 (P_{L_1} and P_{R_1}) and P_2 (P_{L_2} and P_{R_2}) decreased from 0° to 12° and then maintained approximately constant. Specifically, the HPV measured by P_1 (P_{L_1} and P_{R_1}) maintained at about -14 Pa whereas the HPVs measured by P_2 (P_{L_2} and P_{R_2}) maintained at about -4 Pa. One explanation for such characteristics was as follows. As investigated previously in computational flow dynamics simulation, the oscillating amplitude determines the lateral distance between the two vortices shed in one period. When the oscillating amplitude was smaller than 12°, for two vortices shed in one oscillating period, the lateral distance between both of them was less than the width of the robotic fish's head. That is to say, the head

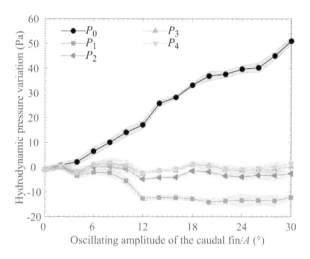

Figure 5.17 Hydrodynamic pressure variation with respect to the oscillating amplitude. The value of P_n equals to the average value of P_{L_n} and P_{R_n} (n=1, 2, 3, and 4).

of the robotic fish blocked the propagation path of the vortices. As a result, the vortices partly burst at the head of the robotic fish. As the oscillating amplitude gradually increased, the shedding position of the vortices gradually moved out of the width of the robotic fish's head. Thus the absolute values of the HPVs measured by P_1 (P_{L_1} and P_{R_1}) and P_2 (P_{L_2} and P_{R_2}) gradually increased. It suggested that the lateral distance between two shedding vortices exceeded the width of the robotic fish's head when the amplitude exceeded 12°. Therefore, the vortices smoothly propagated along both sides of the robotic fish. In addition, because the shedding vortices gradually diffused on the way of propagation, there were pressure differences between the anteroposterior pressure sensors.

5.3.3 Experiment 3: Sensing the Oscillating Frequency of Adjacent Oscillating Caudal Fin

As shown in Figure 5.18, the HPV measured by P_0 increased linearly with the oscillating frequency whereas the HPVs measured by P_1 (P_{L_1} and P_{R_1}) decreased linearly. Besides, the HPVs measured by P_2 (P_{L_2} and P_{R_2}), P_3 (P_{L_3} and P_{R_3}), and P_4 (P_{L_4} and P_{R_4}) maintained around −3 Pa, −1 Pa, and 0, respectively. From the perspective of frequency domain analysis, the frequency components of the HPVs were analyzed, and Figure 5.19 shows a summary statistics of the dominant frequencies

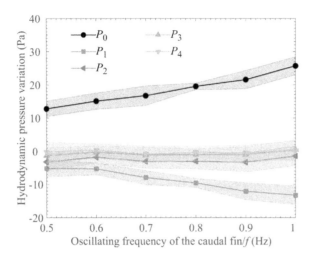

Figure 5.18 Hydrodynamic pressure variation with respect to the oscillating frequency. The value of P_n equals to the average value of P_{L_n} and P_{R_n} (n=1, 2, 3, and 4).

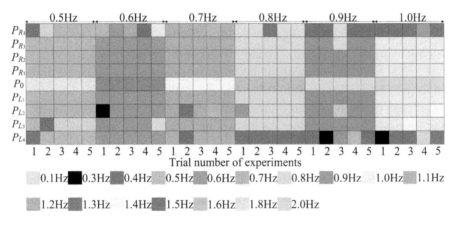

Figure 5.19 A summary statistics of the dominant frequencies in frequency spectrum analysis for the experiment of sensing the oscillating frequency. The color maps represent the dominant frequencies of the hydrodynamic pressure variations.

of the hydrodynamic pressure variations. A careful inspection revealed that, for P_0, the dominant frequency was twice the given oscillating frequency of the caudal fin. For most of the pressure sensors at both sides of the robotic fish, the dominant frequency was equal to the given oscillating frequency of the caudal fin. One possible explanation for such a characteristic was as follows. The upstream oscillating caudal fin shed two vortex rings in one oscillating period, as previously investigated in the computational fluid dynamics simulation. Consequently, the pressure sensor at the tip of the downstream robotic fish probably encountered two vortex rings in one oscillating period. Then the two vortex rings separately propagated along each side of the robotic fish. It was observed that most of the pressure sensors sensed the oscillating frequency precisely. Moreover, comparing with P_0, P_{L_1}, and P_{R_1}, other pressure sensors's accuracies of sensing the oscillating frequency were obviously lower. This is perhaps because the shedding vortices gradually diffused on the way of propagation. As a result, the vortices approaching P_3 (P_{L_3} and P_{R_3}) and P_4 (P_{L_4} and P_{R_4}) were weak and disorganized.

5.3.4 Experiment 4: Sensing the Oscillating Offset of Adjacent Oscillating Caudal Fin

In the experiment of sensing the oscillating offset, the increasing offsets resulted in the deviation of the shedding vortices' propagating direction from the direction of the flow. In this case, the vortex wake generated by the oscillating caudal fin may be dispersed by the flow before propagating downstream. Consequently, the hydrodynamic pressure-bearing area are concentrated on the head of the body of the robotic fish. Thus, we mainly analysed the HPVs measured by P_0, P_{L_1}, and P_{R_1} in order to acquire significant qualitative and quantitative relationships between the HPVs and the oscillating offsets. As shown in Figure 5.20, the curve of the HPV measured by P_0 was symmetrical with respect to the axis $\varphi = 0°$. When the oscillating offset was about $0°$, the HPV reached the maximum value. Besides, when the absolute value of the oscillating offset gradually increased, the scope of the backward reaction gradually deviated from the downstream robotic fish. As a result, the absolute value of the HPV measured by P_0 gradually decreased.

For P_{L_1} and P_{R_1}, the two HPV curves are almost antisymmetrical with respect to the axis $\varphi = 0°$. Take P_{L_1} for example, the HPV decreased then increased with the increasing oscillating offset, and the minimum HPV appeared when the oscillating offset was $5°$. One possible explanation for such a characteristic was as follows. On the one hand, when the oscillating offset increased from $0°$ to $5°$, the vortices propagating to P_{L_1} became relatively more unscathed with less bursting on the head of the robotic fish, thus finally resulting in a biggest increase of the local flow velocity around P_{L_1}. According to Bernoulli's principle, the

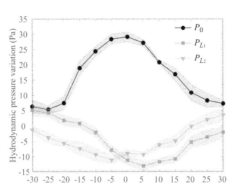

Figure 5.20 Hydrodynamic pressure variation with respect to the oscillating offset.

HPV reached to the minimum value. Then as the oscillating offset continued increasing, to a value of about 8°, the caudal fin gradually turned towards P_{L_1}, as shown in Figure 5.4(c). Consequently, for P_{L_1}, the effect caused by the backward reaction gradually strengthened, whereas the effect caused by the shedding vortices gradually weakened. Therefore, the HPV measured by P_{L_1} relatively increased. Then as the offset continued increasing, the caudal fin gradually deviated from P_{L_1}. Consequently, the scope of the backward reaction gradually departed from P_{L_1}. Simultaneously, as described previously, the vortex wake gradually deviated from the direction of the flow and it may be dispersed by the flow. The above two factors results in weaker and weaker changes in the flow field around P_{L_1}. Thus the absolute value of the HPV measured by P_{L_1} continued decreasing. On the other hand, when the oscillating offset decreased from 0°, for essentially the same reason that the caudal fin gradually departed from P_{L_1}, thus the changes in the flow field around P_{L_1} gradually became weaker, which results in a decreasing absolute value of the HPV. We can also find that, when the oscillating offset were smaller than 12°, the HPV became positive. Such a characteristic may be due to the action of the wave which rebound from the wall of the flume.

5.3.5 Experiment 5: Sensing the Relative Yaw Angle between a Robotic Fish and Its Adjacent Oscillating Caudal Fin

As shown in Figure 5.21(a), the curve of the HPV measured by P_0 was symmetrical with respect to the axis $\alpha = 0°$. When the relative yaw

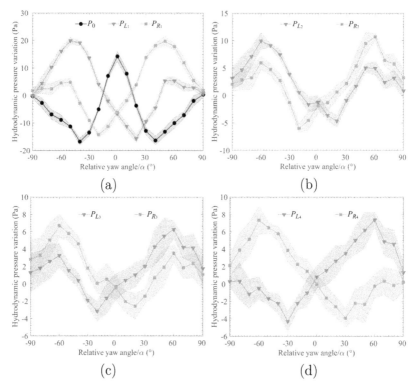

Figure 5.21 Hydrodynamic pressure variation with respect to the relative yaw angle. (a) Hydrodynamic pressure variations measured by P_0, P_{L_1}, and P_{R_1}. (b) Hydrodynamic pressure variations measured by P_{L_2} and P_{R_2}. (c) Hydrodynamic pressure variations measured by P_{L_3} and P_{R_3}. (d) Hydrodynamic pressure variations measured by P_{L_4} and P_{R_4}.

angle was about $0°$, P_0 was located at the region right behind the oscillating caudal fin. As described previously, at this region, the positive effect caused by the backward reaction was strongest, whereas the negative effect caused by the shedding vortices was weakest. As a result, the HPV measured by P_0 reached the maximum value. With the increasing absolute value of relative yaw angle, P_0 gradually moved away from the scope of the backward reaction. Simultaneously, the negative effect caused by the shedding vortices gradually increased. Therefore, the HPV gradually decreased to a negative value and reached to the minimum value when the relative yaw angle was about $\pm 40°$. When the relative yaw angle exceeded $\pm 40°$, because of the continuing expansion of the distance to the caudal fin, both the effects caused by the shedding vortices and the backward reaction decreased. Thus the absolute value of the HPV in the flow field around P_0 gradually decreased. Finally, the HPV measured by P_0 was close to zero.

For P_{L_1} and P_{R_1}, the two curves exhibited the antisymmetry with respect to the axis $\alpha = 0°$. Take P_{L_1} for example, the HPV gradually decreased from $0°$ to $20°$, reaching the minimum value. One possible explanation for such a pressure characteristic was as follows. When the relative yaw angle was about $20°$, P_{L_1} just approached or was fairly close to the propagation path of the shedding vortices. Therefore, the negative effect caused by the shedding vortices reached the peak when the relative yaw angle was around $20°$, thus resulting in a trough value of the HPV. When the relative yaw angle exceed $20°$, the negative effect gradually decreased. Thus the HPV measured by P_{L_1} gradually increased until the yaw angle reached $50°$. It was suggested that the body of the robotic fish prevented the propagation of the shedding vortices when the relative yaw angle exceeded $50°$, thus preventing P_{L_1} from being affected by the change existed in external flow field. As a result, the HPV measured by P_{L_1} gradually decreased to zero. When the relative yaw angle gradually decreased from $0°$ to $-50°$, P_{L_1} gradually approached the scope of the backward reaction. Consequently, the HPV measured by P_{L_1} gradually increased, then reaching the peak. When the relative yaw angle exceeded $-50°$, the absolute value of the HPV measured by P_{L_1} gradually decreased because P_{L_1} gradually departed from the caudal fin. As shown in Figure 5.21(b), for essentially the same reason as P_{L_1} which has been analyzed above, the HPV measured by P_{L_2} exhibited a similar variation tendency. As shown in Figure 5.4(d), P_{L_3} and P_{L_4} are to the right whereas P_{L_1} and P_{L_2} are to the left of the center of mass along the Y-axis. Such a distribution characteristic of pressure sensors results in a characteristic that the variation tendencies of the HPVs measured by P_{L_3} and P_{L_4} were different from those of the HPVs measured by P_{L_1} and P_{L_2}, as shown in Figure 5.21(c) and (d). Specifically, for P_{L_3} and P_{L_4}, when the relative yaw angle gradually decreased from zero, P_{L_3} and P_{L_4} gradually approached and then departed from the shedding vortices, thus resulting in the troughs of the HPVs. When the relative yaw angle gradually increased from zero, the shedding vortices propagating to P_{L_3} and P_{L_4} gradually weakened. Thus the negative effect caused by the vortices gradually decreased. Consequently, the HPVs relatively increased and then reached the second peak with the increasing relative yaw angle. Finally, because of the shadowing effect caused by the body of the robotic fish, the HPVs gradually decreased to zero.

As shown in Figure 5.21, for the pressure sensors at the left side, all the four curves have two peaks and one trough. When the relative yaw angle gradually decreased, the pressure sensors gradually turned to face the caudal fin, that is to say, they gradually faced the scope of the backward reaction, thus resulting in the first peak. A more careful inspection revealed that the value corresponding to each first peak decreased from P_{L_1} to P_{L_4}. Even for P_{L_4}, the first peak disappeared. Besides, the pressure values corresponding to the second peaks maintained approximately constant. One possible explanation for such a characteristic was as follows. On the one hand, because of the distribution characteristics of the pressure sensors, with the increasing relative yaw angles, P_{L_1} and P_{L_2} gradually approached and then departed away from the region right behind the caudal fin, in which the strongest backward reaction existed. However, there always existed certain relative longitudinal distances and lateral distances between P_{L_i} $(i = 3, 4)$ and the caudal fin with the changing relative yaw angle. That is to say, P_{L_3} and P_{L_4} never approached the region right behind the oscillating caudal fin. Moreover, at the region with certain longitudinal distance and lateral distance to the oscillating caudal fin, the positive effect caused by the backward reaction is weaker, whereas the negative effect caused by the shedding vortices is stronger. As a result, the values corresponding to the first peaks of the curves of P_{L_3} and P_{L_4} were both smaller than those of P_{L_1} and P_{L_2}.

Besides, for P_{L_1} and P_{L_2}, when they were located around the region right behind the oscillating caudal fin, the longitudinal distance between P_{L_2} and the caudal fin was always bigger than that between P_{L_1} and the caudal fin. According to the result in our previous study [24], with the increasing relative longitudinal distance to the caudal fin, the HPVs measured by the pressure sensors gradually decreased. Consequently, for P_{L_2}, the value corresponding to the first peak was smaller than that for P_{L_1}. For P_{L_3} and P_{L_4}, when they were located around the region right behind the oscillating caudal fin, the relative longitudinal distance and relative lateral distance between P_{L_4} and the caudal fin were always larger than the two distances between P_{L_3} and the caudal fin. Consequently, comparing with P_{L_3}, the positive effect caused by the reaction was weaker whereas the negative effect caused by the shedding vortices were stronger for P_{L_4}. Thus, for P_{L_4}, the value corresponding to the first peak was smaller than that for P_{L_3}.

On the other hand, as described previously, the strongest vortex intensity appeared on the horizontal symmetrical plane of the caudal fin, then it gradually reduced from middle to both ends of the caudal fin. Considering that comparing with P_{L_1}, P_{L_j} $(j = 2, 3, 4)$ are much farther away from the horizontal symmetrical plane of the caudal fin, therefore, the absolute value of the HPV measured by P_{L_1} was naturally much bigger than those of P_{L_2}, P_{L_3}, and P_{L_4}. Synthesizing the above analyses, the pressure values corresponding to the first peaks gradually decreased from P_{L_1} to P_{L_4}.

As shown in Figure 5.21(c) and (d), there exists big error bars of the hydrodynamic pressure variations measured by P_{L_3}, P_{L_4}, P_{R_3}, and P_{R_4} which are all located at the posterior area of the body of the robot. As shown in Figure 5.7, the shedding vortices gradually diffuse on the way of propagation. Besides, the head of the robotic fish partly blocked the propagation path of the shedding vortices, thus the vortices partly burst at the head of the robotic fish. As a result, the vortices propagating to P_{L_3}, P_{L_4}, P_{R_3}, and P_{R_4} are more diffused. Because there existed unknown disturbances in the flow more or less, which may make the above-mentioned diffusion disorganized and meanwhile affect the laminarity of the flow. All the above-mentioned factors may result in an instable and irregular flow field around P_{L_3}, P_{L_4}, P_{R_3}, and P_{R_4}. Consequently, the error bar of the hydrodynamic pressure variations measured by P_{L_3}, P_{L_4}, P_{R_3}, and P_{R_4} are big.

Figure 5.22(a) shows the right-and-left hydrodynamic pressure difference between two corresponding pressure sensors. All the curves are antisymmetrical with respect to $\alpha = 0°$. Most of the curves exhibited obvious sine-like characteristics. Besides, the variation tendency of the hydrodynamic pressure difference between P_{L_k} and P_{R_k} $(k = 1, 2)$ was opposite to that between P_{L_m} and P_{R_m} $(m = 3, 4)$. As shown in Figure 5.22(b), the overall hydrodynamic pressure difference between left and right sides exhibited a sine-like characteristic. Through curve fitting, the overall difference can be expressed by an equation taking the form:

$$V_{P_L - P_R} = 7.198 \sin (0.03388\alpha + 0.1462) \tag{5.1}$$

where $V_{P_L - P_R}$ was the overall hydrodynamic pressure difference between left and right sides, and α was the relative yaw angle. As shown in Figure 5.22(b), there existed a favorable consistency between the experimental data and the fitting curve (black line) ($R^2 = 0.9543$, RMSE $= 1.1998$).

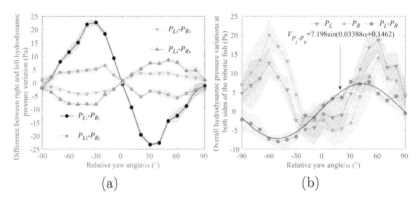

Figure 5.22 Right-and-left hydrodynamic pressure difference with respect to the relative yaw angle. (a) Right-and-left hydrodynamic pressure difference between two corresponding pressure sensors. (b) Overall hydrodynamic pressure variations at both sides of the robotic fish and overall hydrodynamic pressure difference between the two sides. The hydrodynamic pressure variation measured by P_L was equal to the addition of the hydrodynamic pressure variations measured by P_{L_2}, P_{L_3}, and P_{L_4} which are located on the same horizontal section of the body of the robot. The hydrodynamic pressure variation measured by P_R was equal to the addition of the hydrodynamic pressure variations measured by P_{R_2}, P_{R_3}, and P_{R_4}.

5.3.6 Experiment 6: Sensing the Relative Pitch Angle between a Robotic Fish and Its Adjacent Oscillating Caudal Fin

As shown in Figure 5.23, the curve of the HPV measured by P_0 exhibited the symmetry with respect to the axis $\beta = 0°$. When the relative pitch angle was $0°$, P_0 was on the horizontal symmetrical plane of the caudal fin where the vortex intensity was strongest. As a result, the HPV measured by P_0 reached the maximum value when the relative pitch angle was $0°$. As the absolute value of relative pitch angle gradually increased, P_0 gradually moved to both ends of the caudal fin. Thus the vortex intensity around P_0 gradually decreased. As a result, the absolute value of HPV gradually decreased.

As shown in Figure 5.4(e), P_{L_1} and P_{R_1} are beneath whereas P_{L_2}, P_{L_3}, P_{L_4}, P_{R_2}, P_{R_3}, and P_{L_4} are above the center of mass along the Z-axis. Moreover, P_{L_1}, P_{L_2}, P_{R_1}, and P_{R_2} are to the left whereas P_{L_3}, P_{L_4}, P_{R_3}, and P_{L_4} are to the right of the center of mass along the Y-axis. Such distribution characteristics of pressure sensors accounted for the change characteristics of the positions of the pressure sensors when the

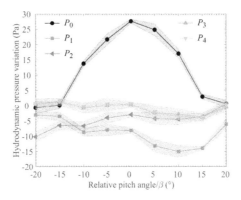

Figure 5.23 Hydrodynamic pressure variation with respect to the relative pitch angle. The value of P_n equals to the average value of P_{L_n} and P_{R_n} (n = 1, 2, 3, and 4).

pitch angle of the robotic fish changed. Specifically, from $-20°$ to $10°$, P_1 (P_{L_1} and P_{R_1}) gradually approached the region with the strongest vortex intensity. Then they gradually moved away from the region when the relative pitch angle exceeded $10°$. For this reason, there appeared a trough value in the HPV curve of P_1 when the relative pitch angle was about $10°$. With the increasing relative pitch angle, P_2 (P_{L_2} and P_{R_2}) gradually moved away from the region with the strongest vortex intensity, whereas P_3 (P_{L_3} and P_{R_3}) and P_4 (P_{L_4} and P_{R_4}) gradually approached the region. Consequently, the HPVs measured by P_2 (P_{L_2} and P_{R_2}) increased progressively, whereas the HPVs measured by P_3 (P_{L_3} and P_{R_3}) and P_4 (P_{L_4} and P_{R_4}) gradually decreased. However, the HPVs measured by P_3 (P_{L_3} and P_{R_3}) and P_4 (P_{L_4} and P_{R_4}) were not obvious perhaps because the vortices gradually diffused before propagating to P_{L_3}, P_{R_3}, P_{L_4}, and P_{R_4}.

5.3.7 Experiment 7: Sensing the Relative Roll Angle between a Robotic Fish and Its Adjacent Oscillating Caudal Fin

As shown in Figure 5.4(f), viewing along the Y-axis, P_0 coincides with the center of mass. Besides, P_{L_1} and P_{R_1} are beneath whereas P_{L_2}, P_{L_3}, P_{L_4}, P_{R_2}, P_{R_3}, and P_{R_4} are all above the center of mass along the Z-axis. Thus the position of P_0 remained unchanged with the change of the relative roll angle. Consequently, as shown in Figure 5.24(a), the HPV measured by P_0 maintained at about 27 Pa. For two corresponding pressure sensors at the left and right sides, the HPVs exhibited the

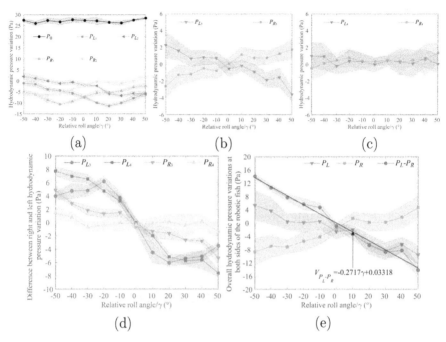

Figure 5.24 Hydrodynamic pressure variation with respect to the relative roll angle. (a) Hydrodynamic pressure variations measured by P_0, P_{L_1}, P_{L_2}, P_{R_1}, and P_{R_2}. (b) Hydrodynamic pressure variations measured by P_{L_3}, P_{L_4}, P_{R_3}, and P_{R_4}. (c) Right-and-left hydrodynamic pressure difference with respect to the relative roll angle. (d) Hydrodynamic pressure variations at both sides of the robotic fish and the overall hydrodynamic pressure difference between the two sides. The value of P_L was equal to the addition of the hydrodynamic pressure variations measured by P_{L_2}, P_{L_3}, and P_{L_4} which are located on the same horizontal section of the body of the robot. The value of P_R was equal to the addition of the hydrodynamic pressure variations measured by P_{R_2}, P_{R_3}, and P_{R_4}.

antisymmetry with respect to the axis $\gamma = 0°$. Take the pressure sensors on the left side for examples, with the increasing relative roll angle, P_{L_1} gradually approached and then moved away from the horizontal symmetrical plane of the caudal fin on which the vortex intensity is strongest. Thus there appeared a trough value in the HPV curve of P_{L_1}. For P_{L_2} and P_{L_3}, with the increasing relative roll angle, P_{L_2} and P_{L_3} gradually moved away from and never approached the horizontal symmetrical plane of the caudal fin. Thus the HPVs measured by P_{L_2} and P_{L_3} decreased progressively. For P_{L_4}, the HPV fluctuated around zero. This is perhaps because that the vortices gradually diffused before propagating to P_{L_4}.

As shown in Figure 5.24(c), all the curves are almost centrosymmetric with respect to the point (0°, 0 Pa). The curve of the hydrodynamic pressure difference between P_{L_1} and P_{R_1} had one peak and one trough. It increased then decreased when the relative roll angle was negative. Besides, the right-and-left hydrodynamic pressure difference between P_{L_2} and P_{R_2} decreased progressively with respect to the relative roll angle. So did the difference between P_{L_3} and P_{R_3}. In addition, the right-and-left pressure difference between P_{L_4} and P_{R_4} fluctuated around zero. As shown in Figure 5.24(d), the hydrodynamic pressure addition at the left side decreased progressively whereas the pressure addition at the right side increased progressively with respect to the relative roll angle. Besides, the overall hydrodynamic pressure difference between right and left sides exhibited a linear characteristic. Through curve fitting, the overall hydrodynamic pressure difference can be expressed by a linear equation taking the form:

$$V_{P_L - P_R} = -0.2717\gamma + 0.03318 \qquad (5.2)$$

where $V_{P_L - P_R}$ was the overall hydrodynamic pressure difference between left and right sides, and γ was the relative roll angle. As shown in Figure 5.24(d), there existed a favorable consistency between the experimental data and the fitting curve (black line) ($R^2 = 0.9821$, RMSE $= 1.2840$). The non-zero phase in equation (5.1) and the constant term in equation (5.2) were possibly caused by the relative position error or attitude error between the robotic fish and its adjacent oscillating caudal fin.

5.4 DISCUSSION

5.4.1 Further Discussion on the Simplification of Sensing the Relative States between Two Adjacent Robotic Fish

Boxfish is named after its box-shaped external structure. Differing from other species of fish, the main body of boxfish is enclosed in a rigid carapace which is not streamlined. Specifically, there are unique keels at the edge of the carapace. The external surface between two adjacent keels is either concave or convex [149]. Previous studies have revealed that the keels of boxfish result in forming self-correcting vortices along both sides of the body, and the vortices give the boxfish the ability of overcoming the motion instability in turbulent flow [149–151]. Considering that the robotic fish used perfectly imitates the appearance characteristics of boxfish in nature, it can be inferred that the robotic fish used is also likely to create its own vortices (which specifically refers to the self-correcting vortices), which may disturb the reverse KVS-like vortex wake generated by its caudal fin. As a result, an individual caudal fin of the robotic fish has been used to replace a complete robotic fish at the upstream region. Thus sensing the relative states between two adjacent robotic fish can be simplified to sensing the relative states between an individual caudal fin and its adjacent robotic fish. Such a simplification of the experiments is conditionally reasonable from the perspective of force, as we described in section 2.2.2. Besides, for a fin-propelled freely-swimming robotic fish, its own motion results in a swaying fish body, thus having effects on hydrodynamic pressure sensing, as discussed in [116, 152]. However, the self-motion effects have not been considered in this work. Specifically, in the above experiments, the individual oscillating caudal fin was fixed at a certain position and was only able to oscillate around a shaft. Besides, the downstream robotic fish was also fixed at certain positions and attitudes, without its own motion. In fact, the yawing motion, pitching motion, and rolling motion are due to the joint action of the paired fins and the caudal fin. However, in order to avoid the fins' interference to the vortex wake, fins of the downstream robotic fish were also removed. In conclusion, it can be seen that the above-mentioned simplifications in the experiments have the limitations for completely emulating the local sensing between two adjacent freely-swimming robotic fish. Nevertheless, the experiments are sufficient to investigate the hydrodynamic characteristics of the reverse KVS-like vortex wake generated by a caudal fin, and the efficient perception of relative states in experiments has demonstrated the effectiveness and practicability of ALLS in local sensing.

5.4.2 The Selection of the Relative Longitudinal Distance between the Individual Caudal Fin and the Robotic Fish

Stewart *et al.* have utilized digital particle image velocimetry (DPIV) technology and computational fluid dynamics (CFD) simulation to explore the flow fields produced by two D-shaped cylinders located in tandem [153]. Specifically, one D-shaped cylinder is located at the upstream region while the other one is located at the downstream region. According to the result in [153], the vortex wake generated by the upstream D-shaped cylinder can be interfered by the downstream D-shaped cylinder. And the whole flow field produced by the two D-shaped cylinders varies when the two cylinders' relative distance along the direction of the flow varies. Inspired by the vortex interactions between the two tandem cylinders, it can be inferred that the vortex wake generated by the oscillating caudal fin in our work can also be interfered by the downstream robotic fish. When the relative longitudinal distance between them is within a certain range, the robotic fish is not just a passive observer but it actively changes the vortex wake. Thus the qualitative and quantitative relationships between the HPVs and the relative states may vary when the relative longitudinal distance varies. However, in our present study, we mainly focus on demonstrating the effectiveness and practicability of ALLS in local sensing for adjacent underwater robots. Thus the relative longitudinal distance between the individual caudal fin and the robotic fish in the experiments was fixed at a certain value of 10 cm when the yaw angle, pitch angle, and roll angle of the robotic fish were all 0°. And we think the efficient perception of relative states in experiments has been sufficient to achieve our research purpose.

On the one hand, before conducting experiments, we have characterized the structure of the shedding vortices behind the robotic fish in a low-turbulence flume with the use of CFD simulation. As shown in Figure 5.5(c), when one vortex completely sheds from the oscillating caudal fin, the shedding vortex has propagated to a position which is around 8.5 cm away from the individual oscillating caudal fin in the longitudinal direction (along the Y-axis). In order to investigate the hydrodynamic characteristics of a well formed vortex, the robotic fish should be placed beyond the above-mentioned position. On the other hand, according to our previous experimental result of sensing the longitudinal distance between adjacent robotic fish that, with the increasing relative longitudinal distance to the caudal fin, the shedding vortices gradually diffuse and finally disappear. Thus the HPVs measured by the ALLS of the downstream robotic fish gradually decreased and finally closed to zero [24]. Consequently, in order to acquire significant qualitative and quantitative relationships between the HPVs and the relative states, we have mainly focused on close-range perception. On the above analyses, the longitudinal distance between the individual caudal fin and the robotic fish was fixed at 10 cm when the relative yaw angle, relative pitch angle, and roll angle were all 0° in all the experiments.

5.4.3 Artificial Lateral Line System's Application in Exploring Reverse Kármán Vortex Street and Kármán Vortex Street

KVS is a typical fluid phenomenon which generally formed behind an obstacle when a running flow is obstructed by the obstacle. In recent years, ALLS has been utilized as an efficient tool to investigate the hydrodynamic characteristics of KVS [12, 28, 84, 116, 139].Reverse KVS is also a fluid phenomenon which is typically generated by the propulsive motion of fish. However, the studies of investigating the reverse KVS using ALLS have been little concerned about in the past.

We have used an oscillating caudal fin to generate a reverse KVS-like vortex wake. Basing on the pressure recordings of the ALLS mounted on a robotic fish at the downstream region, we have investigated the HPVs caused by the reverse KVS. In most of the experiments, when the head of the robotic fish oriented to the incoming flow, the HPV measured by P_0 which was mounted at the nose of the robot reached a value up to around 30 Pa. Even the value increases with the increasing oscillating amplitude and frequency of the caudal fin. And such a value is much bigger than that measured in KVS, when comparing to the HPV measured at the nose of the platform in [12]. Besides, the experimental results have revealed that the HPVs right behind an upstream oscillating caudal fin are positive whereas the HPVs are negative beside the caudal fin at the downstream region, which indicates that there is a high pressure zone in the center of the reverse KVS and a low pressure zone outside the reverse KVS. By comparison, there is a low pressure zone in the center of KVS and a high pressure zone outside the KVS, as investigated in [12].

Venturelli, *et al.* have attempted to recognize and characterize KVS by means of a platform with an onboard pressure sensors array based ALLS [84]. And it has been mentioned that, in KVS, at least half of the sensors simultaneously detect the vortex shedding frequency as the dominant frequency. Combining with the experimental results of sensing the oscillating frequency of the caudal fin, we can come to a conclusion that both the KVS and the reverse KVS can be characterized by the shedding frequency of the vortices. Besides, Venturelli, *et al.* have also found that, in KVS, pressure difference between left and right sides of the platform was linearly correlated to the platform' s orientation (also refers to the relative yaw angle) with respect to the oncoming flow. We have also investigated the relationship between the right-and-left hydrodynamic pressure difference and the relative yaw angle of the robotic fish in reverse KVS, as shown in Figure 5.22(b). However, the relationship is not linear but exhibit a sine-like characteristic. Such a difference between the above-mentioned relationships in KVS and reverse KVS may probably result from the different hydrodynamic characteristics of KVS and reverse KVS, and the different geometries of the robotic fish and the platform. Besides, without more experimental results in a more complete parameter space in [84], we cannot determine if the above-investigated relationship will maintain linear when the platform's orientation becomes larger in KVS. However, a more careful inspection has revealed that when the relative yaw angle belong to $[0°,45°]$, the above-mentioned relationship can also be expressed by a linear equation. In this case, the above-investigated relationship is similar in both KVS and reverse KVS.

5.4.4 Artificial Lateral Line System's Potential in Multiple Underwater Vehicles or Robots-Based Underwater Task Execution

In recent years, autonomous underwater vehicle (AUV) and robot have captured more and more attentions because of their broad prospects for exploiting marine resources, developing the marine economy, and protecting the marine ecological environment. Moreover, it has been a tendency to apply large numbers of underwater vehicles or robots in swarms, for the reason that swarm vehicles or robots have the advantage of increasing sensor density with lower system cost, thus providing more excellent applicability and maneuverability in task execution. However, due to the complexity of underwater working environment, local information interaction for a group of underwater vehicles or robots has been a challenging problem. Though visual sensor and acoustic sensor have provided feasible ways for interacting with the surrounding robots or vehicles, they cannot effectively serve their functions in the hostile environments with complex topography and poor light.

Nevertheless, nature has already provided two exotic ways of interaction, which specifically refer to lateral line-based sensing existed in most species of fish and electro-sensing existed in weakly electric fishes (Gymnotid and Mormyrid) [154, 155]. Both of the two ways are able to serve the function of local sensing and meanwhile play important roles in the localization behavior and communication behavior existed in fish school. Recently, inspired by the biological phenomenon of electro-sensing, several artificial electro-sensing systems have been developed for underwater target localization [156], underwater object identification [157], and control of underwater robot groups [158]. Inspired by the biological phenomenons of lateral line-based sensing, several ALLS have also been developed for underwater robot applications as described previously. For an ALLS composed of pressure sensors array and an artificial electrosensory system (AES) composed of alternating electric dipole, they are still able to function normally under the above-mentioned harsh conditions, thus improving the perception ability of underwater robots and vehicles. For this reason, ALLS and AES are likely to be essential complements to the usual sensor systems equipped with underwater vehicles or robots in the near future. Besides, combining this work with our previous works [24, 146], we have comprehensively investigated the ALLS based close-range perception for relative motions, positions and attitudes between a robotic fish and its neighbour in three-dimensional space. Basing on our researches under laboratory conditions, it is promising to apply ALLS based sensing in local information interaction among multiple vehicles or robots, thus improving underwater task execution in the near future.

5.5 CONCLUSIONS AND FUTURE WORK

Multiple experiments were conducted between a bio-inspired robotic fish with a pressure sensors array based ALLS and its adjacent oscillating caudal fin inside a low-turbulence flume, for investigating the hydrodynamic characteristics of the caudal fin-generating reverse KVS-like vortex wake. And by extracting meaningful information from the pressure sensor readings, the HPVs which reflect the relative states between the robotic fish and its neighbor are investigated efficiently. Specifically, the experimental results have demonstrated that the robotic fish is able to sense the hydrodynamic pressure variations caused by the vortex wake using its onboard ALLS, thus sensing: 1) the relative vertical distance to its adjacent oscillating caudal fin, 2) the oscillating amplitude/frequency/offset of its adjacent oscillating caudal fin, and 3) the relative yaw/pitch/roll angle to its adjacent oscillating caudal fin.

In the future work, we will further conduct experiments of sensing the relative motions, positions, and attitudes between two or among three or more freely-swimming bio-inspired robotic fish by using pressure sensors array based ALLS. On the basis of the qualitative and quantitative relationships between the HPVs and the relative states investigated in the experiments, we will conduct researches on multi-robot formation control using the method of ALLS based local sensing. Specifically, considering that a bio-inspired robotic fish is able to use its onboard ALLS to sense the oscillating states of its adjacent oscillating caudal fin. Therefore, it is promising to design controller for enabling a robotic fish to beat with the same amplitude, frequency and offset as its adjacent robotic fish, thus realizing the leader-follower formation control. Besides, the HPVs right behind an upstream oscillating caudal fin are positive whereas the HPVs are negative beside the caudal fin at the downstream region. On the basis of such a distribution characteristic of the HPVs, we can also design controller to realize the diamond formation of multiple robotic fish with onboard ALLSs, which is a typical swimming pattern existed in fish school [159]. In conclusion, it is promising that ALLS will be critical for formation control of a group of underwater robots and vehicles in the future.

Artificial Lateral Line-Based Relative State Estimation for Two Adjacent Boxfish-Like Robots

T HE LATERAL LINE ENABLES fish to efficiently sense the surrounding environment, thus assisting flow-related fish behaviors. Inspired by this phenomenon, varieties of artificial lateral line systems (ALLSs) have been developed and applied to underwater robots. This chapter focuses on using the pressure sensor arrays based ALLS-measured hydrodynamic pressure variations (HPVs) for estimating the relative states between an upstream oscillating fin and a downstream robotic fish. The HPVs and relative states are measured in flume experiments in which the oscillating fin and the robotic fish have been locate with upstream-downstream formation in a flume. The relative states include the relative oscillating frequency, amplitude, and offset of the upstream oscillating fin to the downstream robotic fish, the relative vertical distance, the relative yaw angle, the relative pitch angle, and the relative roll angle between the upstream oscillating fin and the downstream robotic fish. Regression models between the ALLS-measured and the mentioned relative states are investigated, and regression models-based relative state estimations are conducted. Specifically, two criteria are proposed firstly to

DOI: 10.1201/b23027-6

investigate not only the sensitivity of each pressure sensor to the variations of relative state but also the insufficiency and redundancy of the pressure sensors. And thus the pressure sensors used for regression analysis are determined. Then four typical regression methods, including random forest (RF) algorithm, support vector regression, back propagation neural network, and multiple linear regression method are used for establishing regression models between the ALLS-measured HPVs and the relative states. Then regression effects of the four methods are compared and discussed. Finally, the RF-based method, which has the best regression effect, is used to estimate the relative yaw angle and oscillating amplitude using the ALLS-measured HPVs and exhibits excellent estimation performance.

6.1 INTRODUCTION

Previous studies [24, 160] have mainly focused on the experiments, including experimental conditions, the experimental platform, the method of measuring the ALLS data, the visualisation of the ALLS data, and the explanation of data regularities. However, no models between the ALLS data and the relative states have been built. Thus the results can not be directly used for realising the relative state estimation of two adjacent robotic fish. Whereas we have mainly focused on the applications of the ALLS-measured flow variations in estimating the relative states, which is essential to flow-aided formation control of underwater robot group in the future. Specifically, we have focused on establishing a hydrodynamic model which refers to regression model linking the ALLS-measured HPVs of downstream robotic fish to the relative states.

However, only a few works [34, 92, 144, 161] have investigated modeling for hydrodynamic variations and it is always a challenge to precisely characterize the flow variations caused by the vortices. Potential flow theory has provided a promising approach for quantitatively describing the flow variations existing in fish tail-generated vortex wake [144] and surrounding fish body [34], thus establishing the above-mentioned regression model. However, the existing works have only focused on one individual fish or fish robot [34, 144]. The hydrodynamic variations modeling for two or more fish robots are quite different because it is extremely difficult to establish a model as reference, based on flow dynamics theories [37]. In this case, we attempts to investigate a regression model between the ALLS data and the relative states using intelligent algorithms.

Before conducting the regression modeling, two criteria are proposed firstly for investigating the sensitivity of each pressure sensor to the variations of relative state, then the pressure sensors are sorted according to their sensitivities. Based on the order, the insufficiency and redundancy of the pressure sensors are analyzed in detail. Then the reasonable number of pressure sensors used for regression analysis is determined. To the best of our knowledge, no works have investigated the sensitivity of sensors in ALLS, and only a few works have analyzed insufficiency and redundancy for sensor arrays based ALLS [98, 162]. Such an investigation is helpful for critical and final robotic applications, especially for obtaining the best efficiency of the regression models using the least sensors, and thus decreasing the time of data processing when applying the regression model in online estimation.

Besides, because the regression model between the relative state and the flow variations around the robotic fish is unknown. We can not determine which is the best method for investigating the above regression models. So four typical regression methods which have shown great performance in regression analysis of variables are used for establishing the regression model. The methods include random forest (RF) algorithm, support vector regression (SVR), back propagation neural network (BPNN), and multiple linear regression method (REG). And by comparing the regression effects of using the four methods in detail, the RF method has been determined as the best method. Finally, the estimations for two relative states, relative yaw angle and the oscillating amplitude of the upstream robotic fish, have been conducted to verify the effectiveness of the RF method.

The contributions can be concluded as follows:

1) Guiding the application of ALLS in local perception.

Particularly, we investigates how to obtain the relative states by reversely solving the established regression models using the ALL-measured HPVs. We demonstrates the effectiveness of ALLS in not only close-range perception but also relative state estimation for two or more individuals in an underwater robot group, which has always been a challenge. The work also has the possible extension to underwater robot group control.

2) Quantitatively investigating flow-aided multiple relative states sensing between two underwater robots.

We have investigated multiple relative states using ALLS-measured data measured in a large scope of experimental parameter space based on various intelligent algorithms. To the best of our knowledge, few lateral line inspired researches of underwater robots have investigated various states of the robot. And the experiments have mainly conducted with a fairly limited experimental parameter space.

3) Analysing the sensitivity and insufficiency or redundancy of pressure sensors in ALLS.

We have defined the sensitivity of pressure sensor to the variations of relative state, and then investigated insufficiency or redundancy of the number of the sensors used when extracting information from the ALLS-measured data, the above analyses have been rarely conducted before [162].

6.2 EXPERIMENTAL APPROACH

6.2.1 Experimental Description

As shown in Figure 3.4, in the experiments, we have replaced the upstream robotic fish with an individual caudal fin. And the pectoral fins and caudal fin of the downstream robotic fish have been removed. The two objects are located in a flume with a specific velocity of 0.175 m/s, with given relative states. The positions and attitudes of the two objects are controlled by controllable steering engines linked with them. Thus the relative states can be set, and there are 7 relative states including $d_{vertical}$, f, A, ϕ, α, β, and γ in total. For each relative state, we investigate p experimental parameters, and $p=7$, 16, 6, 13, 19, 9, and 11 for $d_{vertical}$, A, f, ϕ, α, β, and γ, respectively. Details of the p experimental parameters mentioned above are defined in Table 6.1. As we have mentioned in detail, we mainly focuses on investigating the

TABLE 6.1 Details of the p Experimental Parameters.

Items	p	Parameters
$d_{vertical}$ (mm)	7	$-45, -30, -15, 0, 15, 30, 45$
A (degree)	16	$\{0, 2, 4, ..., 26, 28, 30\}$
f (Hz)	6	$0.5, 0.6, 0.7, 0.8, 0.9, 1.0$
ϕ (degree)	13	$\{30, -25, -20, ...,20, 25, 30\}$
α (degree)	19	$\{-90, -80, -70, ..., 70, 80, 90\}$
β (degree)	9	$\{-20, -15, -10, ..., 10, 15, 20\}$
γ (degree)	11	$\{-50, -40, -30, ..., 30, 40, 50\}$

sensitivity of each pressure sensor, the insufficiency and redundancy of the pressure sensors, regression modeling, and model-based relative state estimation using the measure experimental data, to avoid the repetition, more details about such a simplification, the control of relative states between the two adjacent objects, the experimental conditions, the experimental platform, and the method of measuring data can be found in [160]. In the following part, we will use the ALLS-measured HPVs to estimate the above-mentioned relative states using the RF algorithm, BPNN, SVR method, and REG method.

6.2.2 Pretreatment of the Data

Because of the hardware deficiency of the pressure sensors and the background noise in the environment, there exist significant fluctuations of the HPVs measured by the pressure sensors. In this case, we have smoothed the raw HPVs data using a Gaussian smoothing window function. Figure 6.1 shows the raw HPVs and the smoothed HPVs data. In each recording, 250 samples of the HPVs are selected for forming the original sample set O used for regression model analysis. Taking the regression analysis for $d_{vertical}$ for example, there are seven experimental parameters of $d_{vertical}$, varying from -45 mm to 45 mm, with an interval of 15 mm. For each experimental parameter, the HPVs recording is repeated five times. The final samples in the original sample set O for $d_{vertical}$ have a size of $7 \times 5 \times 250 = 8750$.

6.2.3 Random Forest for Regression Task

Random forest, proposed by Breiman, L., is a decision-tree-based ensemble learning algorithm which can be used for classification and regression [163]. The random forest is used for regression here. A random forest can be defined as $R = \{T_1(X), \ldots, T_N(X)\}$, consisting of N decision trees which are defined as $T_i(i = 1, \ldots, N)$. $X = \{X(1), \ldots, X(p)\}$ is a p-dimensional state vector. In a regression task, by importing X into the random forest R, N estimated values $\hat{Y}_i(i = 1, \ldots, N)$ of Y are obtained, where $\hat{Y}_i$ indicates the estimated value of Y obtained by the tree T_i. $\hat{Y} = \frac{\hat{Y}_1 + \cdots + \hat{Y}_N}{N}$ is the estimated result of Y using using multiple classifiers in the random forest. The classifiers can learn and predict the results using the inputs. By importing a sample data set $D = \{(X_1, Y_1), \ldots, (X_n, Y_n)\}$ for training the random forest model, a model which links $X_j(j = 1, \ldots, n)$ to Y_j can be obtained [164]. Each sample in the original

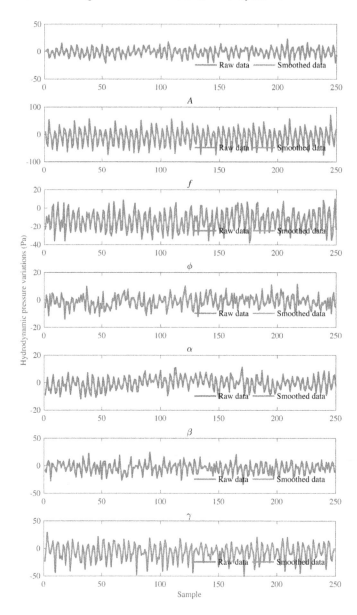

Figure 6.1 Raw data and smoothed pressure sensors-measured hydrodynamic pressure variations in the experiments of investigating $d_{vertical}$, f, A, ϕ, α, β, and γ. The sampling rate of the sensors is 50 Hz.

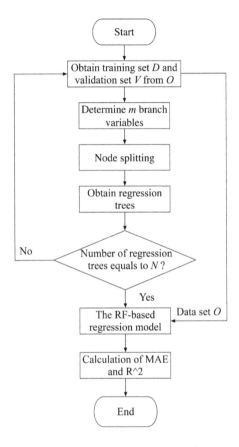

Figure 6.2 Algorithm flow of random forest-based regression analysis.

sample set consists of the HPVs measured by the nine pressure sensors of the ALLS and value of the relative state. Each sample can be defined as $\{S, P_0, P_{L_1}, P_{L_2}, P_{L_3}, P_{L_4}, P_{R_1}, P_{R_2}, P_{R_3}, P_{R_4}\}$, where S indicates value of the relative state, and other variables indicate the ALLS-measured HPVs. For each sample, the state vector X includes nine HPVs measured by the ALLS, and $X = \{P_0, P_{L_1}, P_{L_2}, P_{L_3}, P_{L_4}, P_{R_1}, P_{R_2}, P_{R_3}, P_{R_4}\}$. Y is the relative state S.

Figure 6.2 shows the flow of the random forest for the regression task. It includes four steps described as follows [163, 164].

Step 1: Obtain a bootstrap sample set of size n by randomly sampling with replacement from the original sample set O of size n, as a training set D of the established random forest. The data samples which have not been sampled into the bootstrap sample set are called out-of-bag (OOB) samples and form a validation set V.

Step 2: Randomly select m variables from M variables of the original sample set as branch variables at each node of each regression tree. Then determine the optimum splitting attributes according to the goodness of split criterion. In general, $m = M/3 \pm 1$. If $M < 3$, m is set as 1. The variables of the original sample indicate nine ALLS-measured HPVs, so $M = 1 \sim 9$.

Step 3: Repeat Step 1 and 2 for N times, for obtaining N training sets and N validation sets. Then respectively establish an initial random forest consisting of N regression trees $T_i(i = 1, \ldots, N)$ for the above training sets and validation sets.

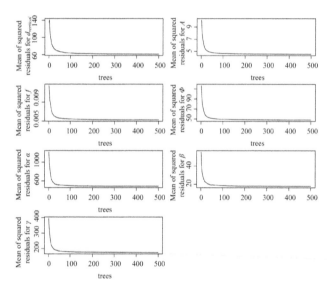

Figure 6.3 Mean of squared residuals with the number of regression trees (N) for $d_{vertical}$, f, A, ϕ, α, β, and γ in Step 3 of random forest based regression task.

Figure 6.3 shows the mean of squared residuals with N for $d_{vertical}$, f, A, ϕ, α, β, and γ. It can be seen that the mean of squared residuals gradually decreases with the increasing N, and it converges to the minimum when N exceeds 400. Here, N is set as 500.

Step 4: Calculate mean absolute error (MAE) and coefficient of determination (R^2) of the estimated results using the original sample set O.

6.2.4 Back Propagation Neural Network (BPNN)

The artificial neural network has been widely used for modeling the relationship between the input signal and output signal. It has exhibited excellent performance in tasks of data classification, data speculation, and pattern identification [165]. BPNN is one of the neural networks which has the widest application. The information in a BPNN propagates forward, and the error propagates backward. The basic theory of BPNN is using gradient descent methods to make the mean square error between the actual output and the expected output least. For a BPNN, it consists of one input layer, one or more hidden layers, and one output layer. In each layer, there are several nodes [166]. The activation function of BPNN is sigmoid function. And we have used L2 regularization (weight decay). A three-layer BPNN has been respectively structured for each experiment, as shown in Figure 6.4. The BPNN has one input layer of

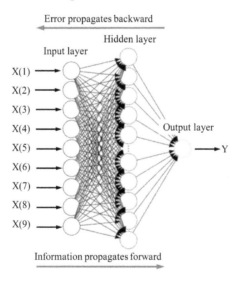

Figure 6.4 Topological structure of the BPNN.

which the number of nodes $p = 9$ and the nine HPVs in each sample are used as the node data. Besides, there is one output layer of which the number of nodes $q = 1$ and the relative state in each sample is used as the node data. Besides, there is one hidden layer of which the number of nodes m is determined as follows.

Figure 6.5 shows the training time and coefficient of determination R^2 with respect to the number of the nodes m in the hidden layer of BPNN for the estimation of β, reported R^2 score is from a training set. It

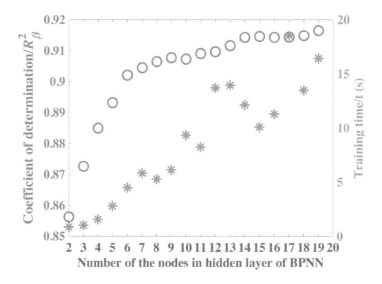

Figure 6.5 Training time (in ∗) and coefficient of determination R^2 (in o) with respect to number of the nodes in the hidden layer of BPNN for the estimation of β. The number of the variables used (M) equals 9. The iterations of BPNN equals 1000.

can be seen that the training time increases with m, and R^2 varies little when the number of nodes exceeds 6. In this case, we have determined m as 6 for the estimation of β. Similarly, m has been determined as 11, 10, 9, 6, 13, 6, and 10 for the estimation of $d_{vertical}$, A, f, ϕ, α, β, and γ, respectively, according to Figure 6.6.

As shown in Figure 6.7, the training time increases with iterations of BPNN, and R^2 varies little when iterations exceeds 400. In this case, iterations of BPNN for α is determined as 400, reported R^2 score is from a training set. Figure 6.8 shows the coefficient of determination R^2 with iterations of BPNN for $d_{vertical}$, A, f, ϕ, α, β, and γ. Similarly, iterations of BPNN for $d_{vertical}$, A, f, ϕ, α, β, and γ are determined as 150, 250, 150, 200, 400, 150, and 300, respectively.

6.2.5 Support Vector Regression (SVR) and Multivariable Linear Regression (REG)

SVR is used to describe regression using support vector methods. It is developed from support vector machine (SVM) by introducing an alternative loss function [167]. SVR with eps-regression and radial basis

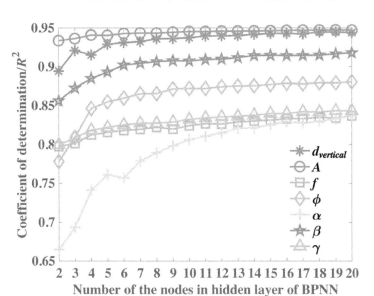

Figure 6.6 Coefficient of determination R^2 with m for all the experiments. Number of the variables used M equals 9.

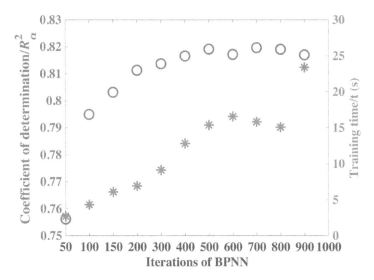

Figure 6.7 Training time (in $*$) and coefficient of determination R^2 (in o) with respect to iterations of BPNN for the estimation of α. The number of the variables used (M) equals 9.

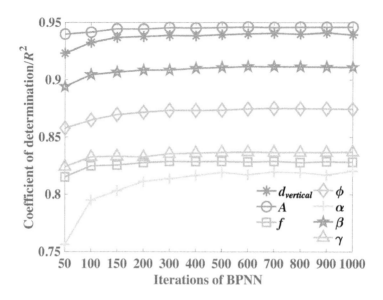

Figure 6.8 Coefficient of determination R^2 with iterations of BPNN for all the experiments. The number of the variables used (M) equals 9.

function as the kernel is used to investigate the regression relationship between the relative state and the ALLS-measured HPVs.

REG refers to investigating a linear regressive relationship between a dependent variable and two or more independent variables. We establish a model for linking the ALLS-measured HPVs to the relative states, as follows:

$$Y = a_0 + \sum a_k X(k) + \varepsilon \qquad (6.1)$$

where Y is the relative state, a_0 is the intercept, a_k is the regression coefficient corresponding to the variable $X(k), k = 1, 2, \cdots, 9$, $X(k)$ indicates the HPVs measured by the nine pressure sensors, and ε is the residual error. F-test is conducted for verifying the rationality of the model. To be specific, the model has the rationality to explain the linear relationship between the dependent variable and the independent variables only when all of the regression coefficients don't equal to zero at the same time.

6.2.6 Sensitivity of the Pressure Sensors-Measured HPVs to the Relative States

In order to study the insufficiency and redundancy of the pressure sensors used for regression analysis, we have first proposed two criteria for measuring the sensitivity of the HPVs measured by the pressure sensors to the variations of relative states. The determination of the two criteria is described as follows.

Step 1: Calculating the variations of the HPVs (ΔHPV) with the variations of the experimental parameters (ΔE) for each pressure sensor. The above experimental parameters include $d_{vertical}$, A, f, ϕ, α, β, and γ.

$$\overline{\Delta HPV_i(k)} = \overline{HPV_{i+1}(k)} - \overline{HPV_i(k)} \tag{6.2}$$

where $i = 1, 2, \cdots, p-1$. p is the number of the experimental parameters, and p=7, 16, 6, 13, 19, 9, and 11 for $d_{vertical}$, A, f, ϕ, α, β, and γ, respectively. $\overline{HPV_i(k)}$ indicates the mean HPVs of 500 samples in the i-th experimental parameter for $X(k)$.

Step 2: Calculating the difference $\overline{\Delta HPV_i(k)}$ between the maximum and the minimum of $\overline{HPV_i(k)}$.

$$mm\overline{HPV_i(k)} = max\overline{HPV_i(k)} - min\overline{HPV_i(k)} \tag{6.3}$$

Step 3: Nondimensionalize the $\overline{\Delta HPV_i(k)}$.

$$\overline{\Delta HPV_i(k)}' = \frac{\overline{\Delta HPV_i(k)}}{mm\overline{HPV_i(k)}} \tag{6.4}$$

Step 4: Calculate the mean of $\overline{\Delta HPV_i(k)}'$ and the mean of $\overline{\Delta HPV_i(k)}$.

$$C_{k_1} = \frac{\sum_{i=1}^{p-1} \overline{\Delta HPV_i(k)}'}{p-1} \tag{6.5}$$

$$C_{k_2} = \frac{\sum_{i=1}^{p-1} \overline{\Delta HPV_i(k)}}{p-1} \tag{6.6}$$

C_{k_1} and C_{k_2} are the two criteria that are respectively used for measuring the sensitivity of $X(k)$ to the variations of relative state. In the experiments, fish head of the downstream robot directly faces to the vortex wake generated by the upstream oscillating caudal fin, so the vortex

TABLE 6.2 C_{k_1} of Each Pressure Sensor in Each Experiment

	$d_{vertical}$	A	f	ϕ	α	β	γ
P_0	0.3165	0.0667	0.2000	0.1633	0.1691	0.2446	0.4535
P_{L_1}	0.3117	0.1109	0.2000	0.1340	0.1268	0.2586	0.1495
P_{L_2}	0.1672	0.2608	0.4971	0.2757	0.1420	0.1713	0.1205
P_{L_3}	0.2676	0.3224	0.3555	0.2394	0.1336	0.2439	0.1000
P_{L_4}	0.4260	0.3172	0.3354	0.2143	0.1113	0.3339	0.2280
P_{R_1}	0.3280	0.0968	0.2025	0.1427	0.1205	0.2199	0.1708
P_{R_2}	0.1723	0.2442	0.5251	0.2713	0.1276	0.1606	0.1167
P_{R_3}	0.3130	0.3174	0.4818	0.2009	0.1346	0.3175	0.1000
P_{R_4}	0.2750	0.3012	0.5028	0.2162	0.1030	0.2462	0.1009

TABLE 6.3 Sorting the Order of the HPVs Measured by the Pressure Sensors According to C_{k_1} (from Biggest to Smallest)

Experiments	Order
$d_{vertical}$	P_{L_4}, P_{R_1}, P_0, P_{R_3}, P_{L_1}, P_{R_4}, P_{L_3}, P_{R_2}, P_{L_2}
A	P_{L_3}, P_{R_3}, P_{L_4}, P_{R_4}, P_{L_2}, P_{R_2}, P_{L_1}, P_{R_1}, P_0
f	P_{R_2}, P_{R_4}, P_{L_2}, P_{R_3}, P_{L_3}, P_{L_4}, P_{R_1}, P_0, P_{L_1}
ϕ	P_{L_2}, P_{R_2}, P_{L_3}, P_{R_4}, P_{L_4}, P_{R_3}, P_0, P_{R_1}, P_{L_1}
α	P_0, P_{L_2}, P_{R_3}, P_{L_3}, P_{R_2}, P_{L_1}, P_{R_1}, P_{L_4}, P_{R_4}
β	P_{L_4}, P_{R_3}, P_{L_1}, P_{R_4}, P_0, P_{L_3}, P_{R_1}, P_{L_2}, P_{R_2}
γ	P_0, P_{L_4}, P_{R_1}, P_{L_1}, P_{L_2}, P_{R_2}, P_{R_4}, P_{L_3}, P_{R_3}

wake has a more significant effect on the fish head. And thus, the HPVs measured by the pressure sensors around the fish head are significantly bigger than HPVs measured by pressure sensors at the posterior. In order to ensure the comparability of different pressure sensors, we have nondimensionalized the $\Delta \overline{HPV_i(k)}$ and then calculate C_{k_1}. On the other hand, we have also calculated calculate C_{k_2} using the $\Delta \overline{HPV_i(k)}$, which has not been nondimensionalized.

Tables 6.2 and 6.4 show C_{k_1} and C_{k_2} of each pressure sensor in each experiments. A bigger value of C_{k_1} or C_{k_2} means a bigger sensitivity. We have sorted the HPVs measured by the pressure sensors from biggest to smallest according to C_{k_1} and C_{k_2}, respectively. Tables 6.3 and 6.5 show the order of the HPVs measured by the pressure sensors from biggest to

TABLE 6.4 C_{k_2} of Each Pressure Sensor in Each Experiment

	$d_{vertical}$	A	f	ϕ	α	β	γ
P_0	11.3143	3.4602	2.5864	3.8769	5.2647	6.9166	0.9537
P_{L_1}	2.3996	1.6968	1.7521	2.4489	4.4564	3.0227	1.6151
P_{L_2}	1.4778	1.3393	1.1334	1.1984	2.0292	1.6177	1.0072
P_{L_3}	0.6341	1.2951	0.8205	1.3811	1.1658	1.6542	0.5840
P_{L_4}	1.0560	1.4444	0.4281	1.1442	1.2445	1.6735	0.2736
P_{R_1}	2.2620	1.4399	1.5111	2.1011	4.1315	2.9065	1.4136
P_{R_2}	1.7180	1.2739	0.8769	1.1659	2.0737	1.9250	0.9149
P_{R_3}	0.7541	1.3676	0.6749	1.1011	1.2411	1.0540	0.4223
P_{R_4}	0.6398	1.2407	0.9156	0.9262	1.2121	1.5360	0.1482

TABLE 6.5 Sorting the Order of the HPVs Measured by the Pressure Sensors According to C_{k_2} (from Biggest to Smallest)

Experiments	Order
$d_{vertical}$	P_0, P_{L_1}, P_{R_1}, P_{R_2}, P_{L_2}, P_{L_4}, P_{R_3}, P_{R_4}, P_{L_3}
A	P_0, P_{L_1}, P_{L_4}, P_{R_1}, P_{R_3}, P_{L_2}, P_{L_3}, P_{R_2}, P_{R_4}
f	P_0, P_{L_1}, P_{R_1}, P_{L_2}, P_{R_4}, P_{R_2}, P_{L_3}, P_{R_3}, P_{L_4}
ϕ	P_0, P_{L_1}, P_{R_1}, P_{L_3}, P_{L_2}, P_{R_2}, P_{L_4}, P_{R_3}, P_{R_4}
α	P_0, P_{L_1}, P_{R_1}, P_{R_2}, P_{L_2}, P_{L_4}, P_{R_3}, P_{R_4}, P_{L_3}
β	P_0, P_{L_1}, P_{R_1}, P_{R_2}, P_{L_4}, P_{L_3}, P_{L_2}, P_{R_4}, P_{R_3}
γ	P_{L_1}, P_{R_1}, P_{L_2}, P_0, P_{R_2}, P_{L_3}, P_{R_3}, P_{L_4}, P_{R_4}

smallest based on the value of C_{k_1} and C_{k_2}. It can be seen that there are significant differences between the two orders of the HPVs.

Then we have successively selected the first (M) sensors in the information-criterion-ordered lists, for establishing the regression model and thus ensuring that the pressure sensors used are not redundant or insufficient. For example, when $M = 3$, we used the first M pressure sensors for regression analysis. The comparisons of regression results corresponding to C_{k_1} and C_{k_2} have been conducted in the following part.

6.2.7 Importance Measurement of the HPVs Measured by Each Pressure Sensor

Importance measurement is conducted for evaluating the importance of the HPV measured by a pressure sensor (variable $X(k)(k = 1,\ldots,p)$

in the state vector X) to the relative state in RF-based analysis. The detailed processes of conducting importance measurement are as follows.

Step 1: Obtain an initial random forest model, including N trees. Then respectively calculate the mean-square error (MSE) of the estimated results using N OOB samples. And the obtained MSEs can be defined as $MSE_i(i = 1, \ldots, N)$, taking the form as

$$MSE_i = \frac{\sum_{j=1}^{n} \left(\hat{Y}_i(j) - Y_i(j) \right)^2}{n} \tag{6.7}$$

where $\hat{Y}_i(j)$ and $Y_i(j)$ indicate the j-th estimated value and actual relative state value in the i-th OOB sample, respectively.

Step 2: Randomly permute $X(k)(k = 1, \ldots, p)$ of the state vector X in the OOB samples, for obtaining new OOB samples. Then calculate the $MSE_i(k)(i = 1, \ldots, N; k = 1, \ldots, p)$ of the estimated results using the new OOB samples. Based on the above processes, a MSE matrix can be obtained, taking the form as

$$\begin{pmatrix} MSE_1(1) & MSE_2(1) & \cdots & MSE_N(1) \\ MSE_1(2) & MSE_2(2) & \cdots & MSE_N(2) \\ \vdots & \vdots & \vdots & \\ MSE_1(p) & MSE_2(p) & \cdots & MSE_N(p) \end{pmatrix} \tag{6.8}$$

Step 3: Calculate the difference between elements in $[MSE_1, \ldots, MSE_N]^T$ and $[MSE_i(1), \ldots, MSE_i(p)]^T$, and the difference can be defined as

$$\Delta MSE_i(k) = MSE_i - MSE_i(k) \tag{6.9}$$

where $i = 1, \ldots, N$, and $k = 1, \ldots, p$.

Step 4: Calculate the importance of variable $X(k)(k = 1, \ldots, p)$. The importance value I_k is defined as

$$I_k = \frac{\sum_{i=1}^{N} \Delta MSE_i(k)}{N \cdot SE_k} \tag{6.10}$$

where SE_k is the standard error of $\Delta MSE_i(k)(i = 1, \ldots, N)$, taking the form as

$$SE_k = \sqrt{\frac{\sum_{i=1}^{N} (\Delta MSE_i(k) - \overline{MSE_k})^2}{N}} \tag{6.11}$$

where $\overline{MSE_k}$ is the mean of $\Delta MSE_i(k)(i = 1, \ldots, N)$. A bigger I_k indicates that $X(k)$ is more important in the regression model.

6.2.8 Evaluation of the Regression Model

Mean absolute error (MAE) and coefficient of determination (R^2) are used for evaluating the accuracy of the above-trained regression model. Smaller MAE and bigger R^2 indicate a more accurate model.

$$MAE = \frac{\sum_{j=1}^{n} |\hat{y}_j - y_j|}{n} \tag{6.12}$$

$$R^2 = 1 - \frac{\sum_{j=1}^{n} (y_j - \hat{y}_j)^2}{\sum_{j=1}^{n} (y_j - \bar{y})^2} \tag{6.13}$$

where $\hat{y}_i$ and y_i indicates the estimated value and actual relative state value in the $i - th$ sample. $\bar{y}$ indicates the mean of $y_j (j = 1, \ldots, n)$.

6.3 RESULTS

6.3.1 Insufficiency and Redundancy of the Pressure Sensors

In investigation of insufficiency and redundancy of the pressure sensors, we aim to find the least number of pressure sensors, with which the accuracy of estimation is big enough. Besides, when the number used for estimation increases, the accuracy of estimation has not increased much more. Such a characteristic means that the pressure sensors is neither insufficient or redundant. Figure 6.9 shows the mean absolute error MAE with the used number of pressure sensors (M). It can be seen that MAE decreases with the increasing M as a whole. Figure 6.10 shows the coefficient of determination R^2 with the used number of sensors. $R^2_{d_{vertical}}$, R^2_A, R^2_f, R^2_A, R^2_ϕ, R^2_α, R^2_β, and R^2_γ indicates R^2 in the experiments of investigating $d_{vertical}$, A, f, ϕ, α, β, and γ. It can be seen that R^2 increases with the number of M as a whole in each experiment when using the four methods. Comparing the MAE and R^2 obtained based on the order of HPVs sorted by C_{k_1} and C_{k_2}, it can be seen that regression models obtained according to order of HPVs sorted by C_{k_2} have better effects. From this perspective, C_{k_2} is more reasonable for measuring the sensitivity of the HPVs measured by pressure sensors to the variations of relative states. In the following part, we mainly focus on the results corresponding to C_{k_2}.

Take the experiment of investigating $d_{vertical}$ for example, a more careful inspection reveals that R^2 and MAE vary little when M exceeds 4. This characteristic demonstrates that 4 pressure sensors which

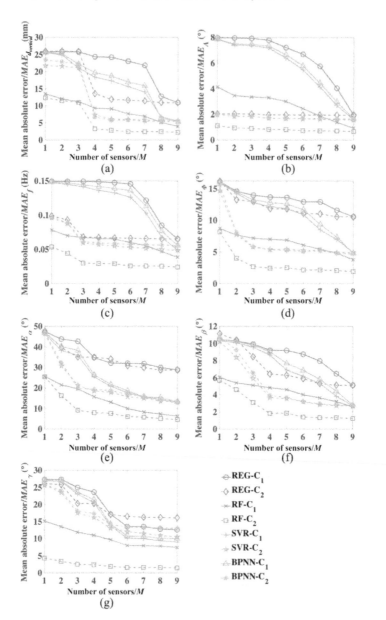

Figure 6.9 Mean absolute error (MAE) with the varied number of sensors (M) when using RF algorithm, BPNN, SVR method, and REG method. Take REG-C_1 for example, it refers to the result obtained by the REG method based on the order of the pressure sensor sorted by C_{k_1}.

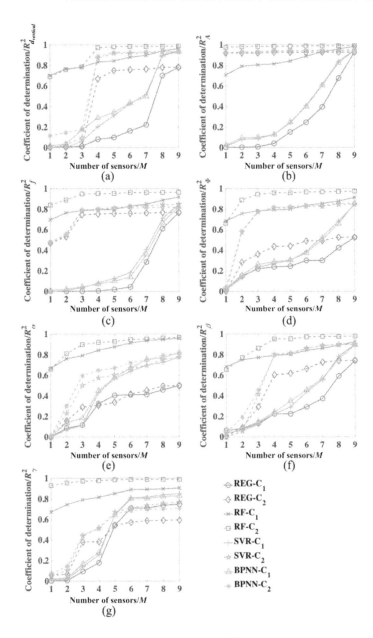

Figure 6.10 Coefficient of determination (R^2) with the varied number of sensors (M) when using RF algorithm, BPNN, SVR method, and REG method.

specifically refer to P_0, P_{L_1}, P_{R_1}, and P_{R_2} here are enough for regression analysis. In other words, five or more pressure sensors are redundant. In this way, the reasonable numbers (defined as M_r) of pressure sensors used for regression analyses of $d_{vertical}$, A, f, ϕ, α, β, and γ are determined as 4, 1, 3, 4, 7, 4, and 5, respectively.

6.3.2 Regression Results Using the Four Methods

In order to have a fair comparison of all of regression methods, similar cross-validations have been conducted when we conducting regression analyses. In each experiment, R^2s obtained by the RF algorithm is bigger than R^2s obtained by BPNN, REG, and SVR. A further inspection has revealed that R^2 obtained by RF has the best performance though M is small. Based on the above analyses, we can conclude that the RF algorithm has better regression effects than the other three methods. Besides, R^2 obtained by the RF algorithm varies smoothly whereas the R^2 obtained by the other three methods have more significant variations with M. Such a characteristic demonstrates that random forest has better noise-resistibility. On the above analyzes, a conclusion can be obtained that RF algorithm has the best regression effect.

Figure 6.11 shows the estimated and actual relative states obtained by the RF algorithm with $M_r =$4, 1, 3, 4, 7, 4, and 5 for the experiment of $d_{vertical}$, A, f, ϕ, α, β, and γ, respectively. On the whole, it can be seen that the established RF based regression models have good performances for describing the relative states. Specifically, the combination of the R^2, MAE, and M_r corresponding to the best regression effect is defined as (R^2, MAE, M), which refers to (0.972, 3.250 mm, 4), (0.975, 1.119°, 1), (0.949, 0.030 Hz, 3), (0.958, 2.467°, 4), (0.952, 5.778°, 7), (0.952, 1.836°, 4), and (0.985, 1.915°, 5) for the experiment of $d_{vertical}$, A, f, ϕ, α, β, and γ, respectively.

Figure 6.12 shows the order of the variables $X(k)$ sorted by the importance I_k. A bigger I_k indicates that the variable $X(k)$ is more important to the relative state in the RF-based regression analysis. The combinations of the M_r most important variables for $d_{vertical}$, A, f, ϕ, α, β, and γ are $(P_0, P_{R_2}, P_{L_2}, P_{L_1})$, (P_0), (P_{R_1}, P_{L_1}, P_0), $(P_{L_1}, P_{R_1}, P_0, P_{L_3})$, $(P_{R_1}, P_{R_2}, P_{L_1}, P_{R_4}, P_0, P_{L_2}, P_{L_4})$, $(P_0, P_{R_2}, P_{L_3}, P_{L_2})$, and $(P_{R_2}, P_{L_2}, P_{L_1}, P_{R_1}, P_{L_3})$, respectively. That is to say, we have determined the M_r pressure sensors used for estimation. A further investigation has revealed that M_r pressure sensors in the above combinations are almost the first M_r pressure sensors in Table 6.5 for each experiment. Such a

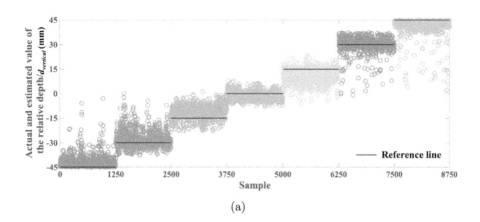

(a)

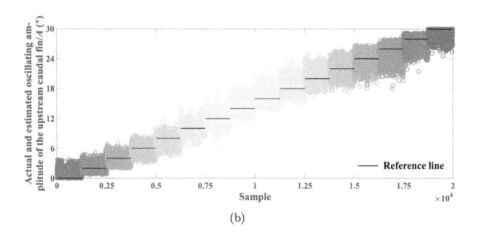

(b)

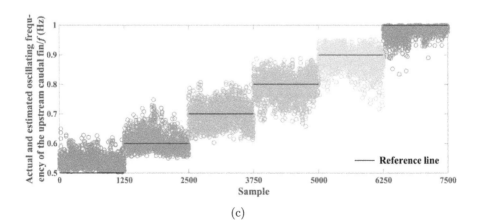

(c)

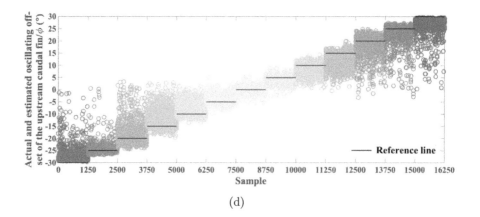

(d)

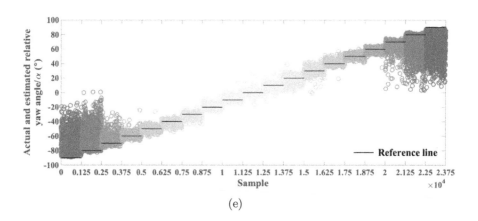

(e)

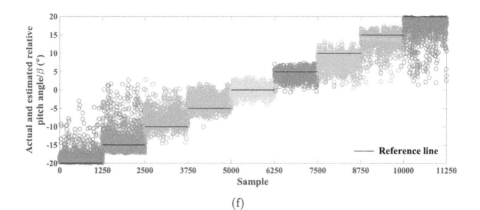

(f)

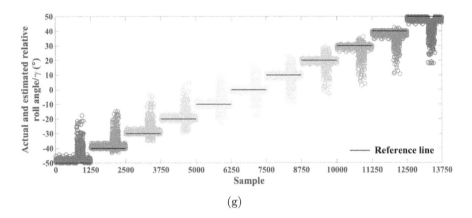

(g)

Figure 6.11 The actual relative states and estimated values obtained by the RF algorithm-based regression model. In each subfigure, the different colors indicate the estimations for specific values of the relative states.

characteristic demonstrates also proves that it is reasonable for sorting the pressure sensors according to C_{k_2} because the first M_r pressure sensors in Table 6.5 happen to be the most important in the regression analyses. Besides, there exist differences between concrete orders of pressure sensors in the above combinations and Table 6.5. Such a characteristic demonstrates that the sensitivity of the pressure sensors-measured HPVs to the relative states is not the same thing as the importance of the HPVs.

6.3.3 Random Forest Algorithm-Based Relative Yaw Angle Estimation and Oscillating Amplitude Estimation

The above work has shown that RF method has the best regression effect. In this part, we validate the effectiveness of RF method in relative state estimation. Considering that the number of the investigated experimental parameters, $d_{vertical}$, A, f, ϕ, α, β, and γ, is 7, 16, 6, 13, 19, 9, and 11, respectively. Here, we select the experiments of investigating A and α for validation works because more experimental parameters have been considered. Specifically, 80% data in the data set O form a training set for training the RF-based models which are different from the models in the section 4.2, and the remaining 20% data form a test set, for validating the effect of the RF-based models in estimating the relative yaw angle and the oscillating amplitude. Figure 6.13 shows the results of RF-based oscillating amplitude and relative yaw angle

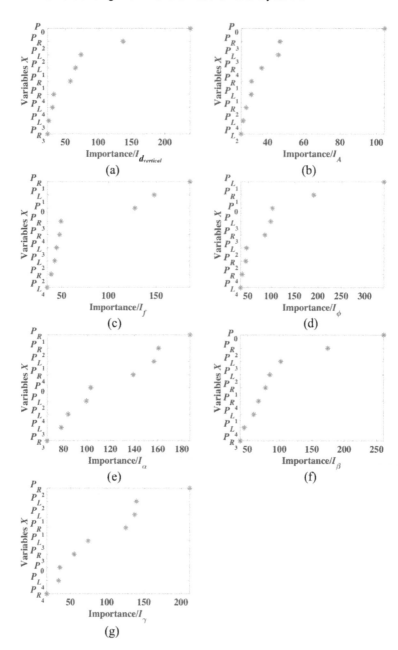

Figure 6.12 Variables $X(k)$ order sorted by the importance I_k.

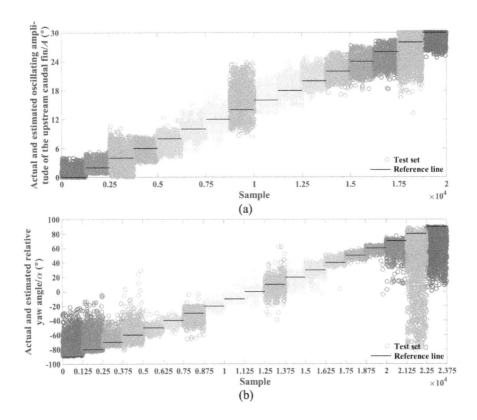

Figure 6.13 Results of RF-based relative yaw angle and oscillating amplitude estimation. (a) Oscillating amplitude estimation ($M_r = 1$, the pressure sensor used is P_0). (b) Relative yaw angle estimation ($M_r = 7$, the pressure sensors used are P_0, P_{L_1}, P_{R_1}, P_{R_2}, P_{L_2}, P_{L_4}, and P_{R_3}). In each subfigure, the different colors indicate the estimations for specific values of the relative states. The grey points mean the estimations of test set.

estimation. R^2 and MAE of the regression model for oscillating amplitude estimation is 0.975 and 1.091°, respectively. And the MAEs of the oscillating amplitude obtained by test set are 2.171°, 3.163°, and 3.211° for $A = 4°, 14°, 28°$, respectively. R^2 and MAE of the regression model for the relative yaw angle is 0.956 and 5.513°, respectively. And the MAEs of the relative yaw angle obtained by test set are 9.284°, 7.506°, 7.694°, and 45.435° for $\alpha = -60°, -30°, 30°, 60°$, respectively. It can be seen that the regression model has a bad performance when the relative yaw angle is big enough. However, it performs well when the relative yaw angle is small. The estimation errors may mainly result from the low noise-signal ratio of the HPVs measured by the pressure sensors. Though the pretreatment of the HPVs has been conducted, the hardware deficiency of the pressure sensors has reduced the estimation accuracy. In this case, improving the ALLS is necessary and urgent for improving the data quality. Besides, further data pretreatment by fusing ALLS data and IMU data may also provide a potential way for improving the estimation accuracy. The above-mentioned will be conducted in the following researches on multiple adjacent freely-swimming robotic fish.

6.4 DISCUSSIONS

6.4.1 Why Have We Focused on Close-Range Sensing?

As shown in Figure 6.14, the relationship between the HPV and the relative lateral distance varies with the different relative longitudinal distances. There is a common qualitative regularity existing in the HPV curves under a short-range relative longitudinal distance. However, when the relative longitudinal distance exceeds a specific value, the above-mentioned qualitative regularity disappears. This may be because the vortex wake generated by the upstream robotic fish scatters and disappears with the increasing propagation distance. To acquire significant regularities between the HPVs and the relative states, we have mainly focused on close-range sensing (local sensing), and the relative longitudinal distance of the two adjacent robotic fish has been fixed at 10 cm in [160].

Such a relative longitudinal distance is relatively small when compared with the size of the robot. So the downstream robotic fish is not just a passive observer, but it actively changes the vortex wake generated by the upstream oscillating caudal fin. Figure 6.15 shows the instantaneous vortex structure by Q-criterion around the robotic fish body and hydrodynamic pressure variations on the surface of the robotic fish body. They are obtained by the computational fluid dynamics simulation software called Fluent. More details about the Fluent based simulation can be found in [160]. As shown in Figure 6.15, the vortices disperse after they propagate over the fish head. As a result, the HPVs measured by the pressure sensors at the posterior of the downstream robotic fish have small variations.

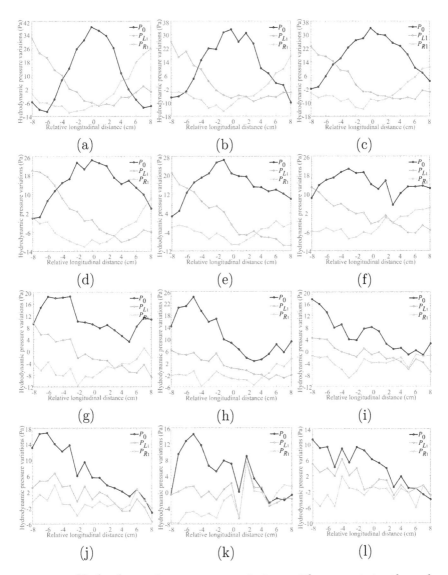

Figure 6.14 Hydrodynamic pressure variation with respect to the relative lateral distance $d_{lateral}$ under different relative longitudinal distances $d_{longitudinal}$. (a) $d_{longitudinal}$=4 cm, (b) $d_{longitudinal}$=6 cm, (c) $d_{longitudinal}$=8 cm, (d) $d_{longitudinal}$=10 cm, (e) $d_{longitudinal}$=12 cm, (f) $d_{longitudinal}$=14 cm, (g) $d_{longitudinal}$=16 cm, (h) $d_{longitudinal}$=18 cm, (i) $d_{longitudinal}$=20 cm, (j) $d_{longitudinal}$=22 cm, (k) $d_{longitudinal}$=24 cm, (l) $d_{longitudinal}$=26 cm.

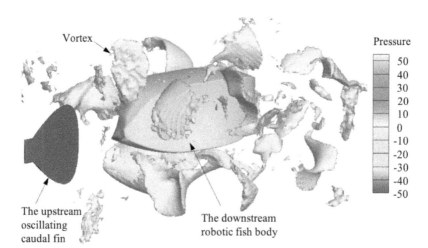

Figure 6.15 Instantaneous vortex structure by Q-criterion around the robotic fish body and hydrodynamic pressure variations on the surface of the robotic fish body.

6.4.2 The Differences between Investigating the Relative Yaw Angle between Two Adjacent Robotic Fish and Investigating the Oscillating Offset of the Upstream Oscillating Caudal Fin

The differences include three parts, such as the following:

First, in the experiments, the upstream oscillating caudal fin and the downstream robotic fish played roles as "vortex signal generating source" and "vortex signal detector", respectively. So, the difference between investigating the relative yaw angle of the downstream robotic fish and investigating the oscillating offset of the upstream oscillating caudal fin was the difference between changing the "vortex signal generating source" and changing "vortex signal detector". Specifically, in the experiment of investigating the oscillating offset, the nonzero oscillating offset resulted in the deviation of the shedding vortices' propagating direction from the direction of the flow. Meanwhile, the scope of the backward reaction also deviated from the robotic fish in the downstream region. However, the downstream robotic fish that served the function of signal detector never changed. In the experiment of investigating the relative yaw angle, the pressure sensors mounted on the downstream robotic fish who served the function of hydrodynamic signal detector changed its orientation to the incoming flow as its attitude changed. Meanwhile, the body of the robotic fish may also cause flow separation [168] around itself and resulted in the self-generating vortices [151]. However, the upstream oscillating caudal fin who served as the hydrodynamic signal generating source never changed.

Second, in the experiment of investigating the oscillating offset of the upstream oscillating caudal fin, because the shedding vortices' propagating direction deviated from the direction of the flow, the vortex wake generated by the oscillating caudal fin may be dispersed by the flow. Thus it may be not as well-formed and stable as the vortex wake generated in the experiment of investigating the relative yaw angle.

Third, for the experiment of investigating the oscillating offset of the upstream oscillating caudal fin, the hydrodynamic pressure-bearing area of the downstream robotic fish was concentrated on the head of its body. Moreover, with the increasing oscillating offset, the above-mentioned area gradually decreased. This may result from the deviation of the vortex wake, as described in the second point. By comparison, the hydrodynamic pressure-bearing area can be almost distributed on the whole body of the robot with the changing orientation of the robot to the incoming flow in the experiment of investigating the relative yaw angle. The above-mentioned differences have resulted in the different qualitative and quantitative regularities of hydrodynamic pressure variations with respect to the experimental parameters. Specifically, in the experiment of investigating the oscillating offset, the HPVs measured by P_0, P_{L_1}, and P_{R_1} change apparently with the oscillating offsets, whereas the changes of the HPVs measured by other pressure sensors are not apparent with the changing oscillating offsets. By comparison, in the experiment of investigating the relative yaw angle, the HPVs measured by all the pressure sensors have apparent changes with the changing relative yaw angles.

6.5 CONCLUSIONS AND FUTURE WORK

We have investigated the regression model, which links the ALLS-measured HPVs to the relative states between two adjacent robotic fish with leader-follower formation in a flume. Two criteria are proposed firstly for investigating not only the sensitivity of each pressure sensor to the variations of relative state variations, but also the insufficiency and redundancy of the pressure sensors used for regression analysis. Then four methods, including RF algorithm, BPNN, SVR, and REG, are used for establishing the regression model. Comparisons of the effects of the regression effects of using the four methods have been conducted, for determining the best method. The results show that the RF algorithm has the best excellent performance in estimating the relative states using the ALLS-measured HPVs. And RF-based regression models have been established for predicting the relative yaw angle and the oscillating amplitude of the upstream robotic fish with small errors. Moreover, a further discussion for the flume experiments has been conducted in detail.

For different underwater vehicles like propeller-actuated autonomous underwater vehicle (AUV) and remotely operated vehicle (ROV), the flow variations caused by the propeller and vehicle motions are essentially different from fin-actuated robot. We have selected a specifically designed mechatronic system (robotic fish) for investigating artificial lateral line system based relative state estimation for underwater robots group. The regression methods used could be used in other underwater mechatronic systems.

In the future, we will conduct an online estimation of the relative states between two adjacent freely-swimming robotic fish using the RF-based regression model. Free motions of the robotic fish result in rhythmical oscillations of the fish body, which may have significant effects on the ALLS-measured HPVs [36]. So the qualitative and quantitative relationships between the HPVs and the relative states may significantly differ from those we have investigated and [24, 160]. Relative state estimation for two adjacent freely-swimming robotic fish would be more challenging.

CHAPTER 7

SUMMARY

L ATERAL LINE SYSTEM is a flow-responsive organ system with which fish can effectively sense the surrounding flow field, thus serving functions in flow-aided fish behaviors. Inspired by such a biological characteristic, artificial lateral line systems (ALLSs) have been developed for promoting technological innovations of underwater robots. This book focuses on bionic sensing with artificial lateral line systems for fish-like underwater robot.

In this book, we first introduced two fish-like underwater robots, including a multiple fins-actuated robotic fish and a caudal fin-actuated robotic fish with barycenter regulating mechanism. Then we studied how a robotic fish uses its onboard pressure sensor arrays based-ALLS to estimate its trajectory in multiple locomotions, including rectilinear motion, turning motion, ascending motion, and spiral motion, thus realizing the online localization. Finally, we explored the ALLS-based relative position and attitude perception between two robotic fish in a leader-follower formation. Four regression methods, including multiple linear regression method, support vector regression, back propagation neural network, and random forest method, were used to evaluate the relative positions or attitudes using the ALLS data. The results indicated that the robotic fish can evaluated its motion parameters using the ALLS-measured pressure variations surrounding the fish robot body with small errors, and further evaluating the trajectory efficiently using the evaluated motion parameters. The research on ALLS-based local sensing between two adjacent fish robots extended the current research focus from one individual underwater robot to two robots with a formation.

DOI: 10.1201/b23027-7

Bibliography

[1] S. Coombs, H. Bleckmann, R. R. Fay, and A. N. Popper. *The lateral line system*. Springer New York, New York, 2014.

[2] J. Mogdans and H. Bleckmann. Coping with flow: behavior, neurophysiology and modeling of the fish lateral line system. *Biological Cybernetics*, 106(11):627–642, 2012.

[3] J. C. Montgomery, C. F. Baker, and A. G. Carton. The lateral line can mediate rheotaxis in fish. *Nature*, 389(6654):960–963, 1997.

[4] H. Bleckmann. *Role of the lateral line in fish behaviour*. Springer US, 1986.

[5] S. Tan. Underwater artificial lateral line flow sensors. *Microsystem Technologies*, 20(12):2123–2136, 2014.

[6] G. Liu, A. Wang, X. Wang, and P. Liu. A review of artificial lateral line in sensor fabrication and bionic applications for robot fish. *Applied Bionics and Biomechanics*, 2016(5):1–15, 2016.

[7] Y. Yang, N. Nguyen, N. Chen, M. Lockwood, and D. L. Jones. Artificial lateral line with biomimetic neuromasts to emulate fish sensing. *Bioinspiration & Biomimetics*, 5(1):16001, 2010.

[8] A. Ahrari, H. Lei, M. Sharif, K. Deb, and X. Tan. Reliable underwater dipole source characterization in three-dimensional space by an optimally designed artificial lateral line system. *Bioinspiration & Biomimetics*, 12(3):036010, 2017.

[9] L. H. Boulogne, B. J. Wolf, M. A. Wiering, and S. M. van Netten. Performance of neural networks for localizing moving objects with an artificial lateral line. *Bioinspiration & Biomimetics*, 12(5):056009, 2017.

[10] X. Zheng, Y. Zhang, M. Ji, Y. Liu, X. Lin, J. Qiu, and G. Liu. Underwater positioning based on an artificial lateral line and a generalized regression neural network. *Journal of Bionic Engineering*, 15(5):883–893, 2018.

[11] M. Asadnia, A. G. Kottapalli, J. Miao, M. E. Warkiani, and M. S. Triantafyllou. Artificial fish skin of self-powered micro-electromechanical systems hair cells for sensing hydrodynamic flow phenomena. *Journal of the Royal Society Interface*, 12(111):20150322, 2015.

[12] L. D. Chambers, O. Akanyeti, R. Venturelli, J. Ježov, J. Brown, M. Kruusmaa, P. Fiorini, and W. M. Megill. A fish perspective: detecting flow features while moving using an artificial lateral line in steady and unsteady flow. *Journal of the Royal Society Interface*, 11(99):20140467, 2014.

[13] N. Strokina, J. K. Kämäräinen, J. A. Tuhtan, J. F. Fuentes-Pérez, and M. Kruusmaa. Joint estimation of bulk flow velocity and angle using a lateral line probe. *IEEE Transactions on Instrumentation and Measurement*, 65(3):601–613, 2016.

[14] E. Kanhere, N. Wang, A. G. Kottapalli, M. Asadnia, V. Subramaniam, J. Miao, and M. Triantafyllou. Crocodile-inspired dome-shaped pressure receptors for passive hydrodynamic sensing. *Bioinspiration & Biomimetics*, 11(5):056007, 2016.

[15] K. Chen, J. A. Tuhtan, J. F. Fuentes-Pérez, G. Toming, M. Musall, N. Strokina, J. K. Kämäräinen, and M. Kruusmaa. Estimation of flow turbulence metrics with a lateral line probe and regression. *IEEE Transactions on Instrumentation and Measurement*, 66(4):651–660, 2017.

[16] Y. Jiang, Z. Ma, J. Fu, and D. Zhang. Development of a flexible artificial lateral line canal system for hydrodynamic pressure detection. *Sensors*, 17(6):1220, 2017.

[17] B. J. Wolf, J. Morton, W. Macpherson, and S. N. Van. Bio-inspired all-optical artificial neuromast for 2d flow sensing. *Bioinspiration & Biomimetics*, 13(2):026013, 2018.

[18] G. Liu, M. Wang, A. Wang, S. Wang, T. Yang, R. Malekian, and Z. Li. Research on flow field perception based on artificial lateral line sensor system. *Sensors*, 18(3):838, 2018.

[19] L. DeVries, F. D. Lagor, H. Lei, X. Tan, and D. A Paley. Distributed flow estimation and closed-loop control of an underwater vehicle with a multi-modal artificial lateral line. *Bioinspiration & Biomimetics*, 10(2):025002, 2015.

[20] W. K. Yen and J. Guo. Phase controller for a robotic fish to follow an oscillating source. *Ocean Engineering*, 161:77–87, 2018.

[21] W. K. Yen, D. M. Sierra, and J. Guo. Controlling a robotic fish to swim along a wall using hydrodynamic pressure feedback. *IEEE Journal of Oceanic Engineering*, 43(2):369–380, 2018.

[22] W. Wang, Y. Li, X. Zhang, C. Wang, S. Chen, and G. Xie. Speed evaluation of a freely swimming robotic fish with an artificial lateral line. In *IEEE International Conference on Robotics and Automation*, pages 4737–4742, Stockholm, Sweden, May 2016.

[23] X. Zheng, C. Wang, R. Fan, and G. Xie. Artificial lateral line based local sensing between two adjacent robotic fish. *Bioinspiration & Biomimetics*, 13(1):016002, 2018.

[24] W. Wang, X. Zhang, J. Zhao, and G. Xie. Sensing the neighboring robot by the artificial lateral line of a bio-inspired robotic fish. In *IEEE/RSJ International Conference on Intelligent Robots and Systems*, pages 1565–1570, Hamburg, Germany, September 2015.

[25] Y. Xu and K. Mohseni. Fish lateral line inspired hydrodynamic feedforward control for autonomous underwater vehicles. In *IEEE/RSJ International Conference on Intelligent Robots and Systems*, pages 3565–3870, Tokyo, Japan, November 2013.

[26] T. Salumäe, I. Ranó, O. Akanyeti, and M. Kruusmaa. Against the flow: A braitenberg controller for a fish robot. In *IEEE International Conference on Robotics and Automation*, pages 4210–4215, Saint Paul, USA, May 2012.

[27] J. Ježov, O. Akanyeti, L. D. Chambers, and M. Kruusmaa. Sensing oscillations in unsteady flow for better robotic swimming efficiency. In *IEEE International Conference on Systems, Man, Cybernetics*, pages 91–96, Seoul, Korea, October 2012.

[28] T. Salumäe and M. Kruusmaa. Flow-relative control of an underwater robot. *Proceedings of the Royal Society A: Mathematical, Physical and Engineering Sciences*, 469(2153):20120671, 2013.

[29] A. Gao and M. Triantafyllou. Bio-inspired pressure sensing for active yaw control of underwater vehicles. In *2012 Oceans*, pages 1–7, Hampton Roads, VA, USA, October 2012.

[30] J. Dusek, A. G. P. Kottapalli, M. E. Woo, M. Asadnia, J. Miao, J. H. Lang, and M. S. Triantafyllou. Development and testing of bio-inspired microelectromechanical pressure sensor arrays for increased situational awareness for marine vehicles. *Smart Materials and Structures*, 22(22):014002, 2013.

[31] M. Asadnia, A. G. Kottapalli, R. Haghighi, A. Cloitre, P. V. Alvarado, J. Miao, and M. Triantafyllou. Mems sensors for assessing flow-related control of an underwater biomimetic robotic stingray. *Bioinspiration & Biomimetics*, 10(3):036008, 2015.

[32] A. G. P. Kottapalli, M. Asadnia, Z. Shen, V. Subramaniam, J. Miao, and M. Triantafyllou. Mems artificial neuromast arrays for hydrodynamic control of soft-robots. In *IEEE International Conference on Nano/micro Engineered and Molecular Systems*, Sendai, Japan, April 2016.

[33] M. Asadnia, A. G. P. Kottapalli, Z. Shen, J. Miao, and M. Triantafyllou. Flexible and surface-mountable piezoelectric sensor arrays for underwater sensing in marine vehicles. *IEEE Sensors Journal*, 13(10):3918–3925, 2013.

[34] X. Zheng, W. Wang, M. Xiong, and G. Xie. Online state estimation of a fin-actuated underwater robot using artificial lateral line system. *IEEE Transactions on Robotics*, 2020.

[35] X. Zheng, C. Wang, R. Fan, and G. Xie. Artificial lateral line based local sensing between two adjacent robotic fish. *Bioinspiration & Biomimetics*, 13(1):016002, 2017.

[36] X. Zheng, M. Wang, J. Zheng, R. Tian, M. Xiong, and G. Xie. Artificial lateral line based longitudinal separation sensing for two swimming robotic fish with leader-follower formation. In *2019 IEEE/RSJ International Conference on Intelligent Robots and Systems (IROS)*, pages 2539–2544. IEEE, 2019.

[37] X. Zheng, M. Xiong, and G. Xie. Data-driven modeling for superficial hydrodynamic pressure variations of two swimming robotic fish with leader-follower formation. In *2019 IEEE International Conference on Systems, Man and Cybernetics (SMC)*, pages 4331–4336. IEEE, 2019.

[38] M. S. Triantafyllou and G. S. Triantafyllou. An efficient swimming machine. *Scientific American*, 272(3):64–70, 1995.

[39] J. Yu, L. Liu, L. Wang, M. Tan, and D. Xu. Turning control of a multilink biomimetic robotic fish. *IEEE Transactions on Robotics*, 24(1):201–206, 2008.

[40] J. Liang, T. Wang, and L. Wen. Development of a two-joint robotic fish for real-world exploration. *Journal of Field Robotics*, 28(1):70–79, 2011.

[41] W. Wang and G. Xie. Online high-precision probabilistic localization of robotic fish using visual and inertial cues. *IEEE Transactions on Industrial Electronics*, 62(2):1113–1124, 2014.

[42] J. Yu, M. Wang, M. Tan, and J. Zhang. Three-dimensional swimming. *IEEE Robotics & Automation Magazine*, 18(4):47–58, 2011.

[43] A. Crespi, D. Lachat, A. Pasquier, and A. J. Ijspeert. Controlling swimming and crawling in a fish robot using a central pattern generator. *Autonomous Robots*, 25(1-2):3–13, 2008.

[44] K. Seo, S.-J. Chung, and J.-J. E. Slotine. Cpg-based control of a turtle-like underwater vehicle. *Autonomous Robots*, 28(3):247–269, 2010.

[45] A. J. Ijspeert, A. Crespi, D. Ryczko, and J.-M. Cabelguen. From swimming to walking with a salamander robot driven by a spinal cord model. *Science*, 315(5817):1416–1420, 2007.

[46] J. Mogdans and H. Bleckmann. Coping with flow: behavior, neurophysiology and modeling of the fish lateral line system. *Biological Cybernetics*, 106(11-12):627–642, 2012.

[47] R. G. Northcutt. The phylogenetic distribution and innervation of craniate mechanoreceptive lateral lines. In *The mechanosensory lateral line*, pages 17–78. Springer, 1989.

[48] K. P. Maruska. Morphology of the mechanosensory lateral line system in elasmobranch fishes: ecological and behavioral considerations. *Environmental Biology of Fishes*, 60(1-3):47–75, 2001.

[49] G. Liu, A. Wang, X. Wang, and P. Liu. A review of artificial lateral line in sensor fabrication and bionic applications for robot fish. *Applied Bionics and Biomechanics*, 2016, 2016.

[50] S. Coombs, J. Janssen, and J. F. Webb. Diversity of lateral line systems: evolutionary and functional considerations. In *Sensory biology of aquatic animals*, pages 553–593. Springer, 1988.

[51] H. Münz. Morphology and innervation of the lateral line system insarotherodon niloticus (l.)(cichlidae, teleostei). *Zoomorphologie*, 93(1):73–86, 1979.

[52] T. Shizhe. Underwater artificial lateral line flow sensors. *Microsystem Technologies*, 20(12):2123–2136, 2014.

[53] S. M. van Netten. Hydrodynamic detection by cupulae in a lateral line canal: functional relations between physics and physiology. *Biological Cybernetics*, 94(1):67–85, 2006.

[54] M. J. McHenry, J. A. Strother, and S. M. Van Netten. Mechanical filtering by the boundary layer and fluid–structure interaction in the superficial neuromast of the fish lateral line system. *Journal of Comparative Physiology A*, 194(9):795, 2008.

[55] Z. Fan, J. Chen, J. Zou, D. Bullen, C. Liu, and F. Delcomyn. Design and fabrication of artificial lateral line flow sensors. *Journal of Micromechanics & Microengineering*, 12(5):655–661, 2002.

[56] N. Chen, C. Tucker, J. M. Engel, Y. Yang, and L. Chang. Design and characterization of artificial haircell sensor for flow sensing with ultrahigh velocity and angular sensitivity. *Journal of Microelectromechanical Systems*, 16(5):999–1014, 2007.

[57] M. E. Mcconney, N. Chen, D. Lu, H. A. Hu, S. Coombs, C. Liu, and V. V. Tsukruk. Biologically inspired design of hydrogel-capped hair sensors for enhanced underwater flow detection. *Soft Matter*, 5(2):292–295, 2009.

[58] A. Qualtieri, F. Rizzi, M. T. Todaro, A. Passaseo, R. Cingolani, and M. De Vittorio. Stress-driven A1N cantilever-based flow sensor for fish lateral line system. *Microelectronic Engineering*, 88(8):2376–2378.

[59] A. Qualtieri, F. Rizzi, G. Epifani, A. Ernits, M. Kruusmaa, and M. De Vittorio. Parylene-coated bioinspired artificial hair cell for liquid flow sensing. *Microelectronic Engineering*, 98(Complete):516–519.

[60] A. G. P. Kottapalli, M. Asadnia, J. Miao, and M. Triantafyllou. Touch at a distance sensing: lateral-line inspired mems flow sensors. *Bioinspiration & Biomimetics*, 9(4):046011.

[61] A. G. P. Kottapalli, M. Bora, M. Asadnia, J. Miao, S. S. Venkatraman, and M. Triantafyllou. Nanofibril scaffold assisted mems artificial hydrogel neuromasts for enhanced sensitivity flow sensing. *Scientific Reports*, 6(1):1–12, 2016.

[62] V. I. Fernandez, S. M. Hou, F. S. Hover, J. H. Lang, and M. S. Triantafyllou. Lateral-line inspired mems-array pressure sensing for passive underwater navigation. Technical report, Massachusetts Institute of Technology. Sea Grant College Program, 2007.

[63] F. M. Yaul, V. Bulovic, and J. H. Lang. A flexible underwater pressure sensor array using a conductive elastomer strain gauge. *Journal of Microelectromechanical Systems*, 21(4):897–907, 2012.

[64] A. G. P. Kottapalli, M. Asadnia, J. M. Miao, G. Barbastathis, and M. S. Triantafyllou. A flexible liquid crystal polymer mems pressure sensor array for fish-like underwater sensing. *Smart Materials and Structures*, 21(11):115030, 2012.

[65] J. Chen, Z. Fan, J. Zou, J. Engel, and L. Chang. Two-dimensional micromachined flow sensor array for fluid mechanics studies. *Journal of Aerospace Engineering*, 16(2):85–97, 2003.

[66] Y. Yang, N. Chen, C. Tucker, J. Engel, S. Pandya, and C. Liu. From artificial hair cell sensor to artificial lateral line system: development and application. In *2007 IEEE 20th International Conference on Micro Electro Mechanical Systems (MEMS)*, pages 577–580. IEEE, 2007.

[67] N. Chen, J. Chen, J. Engel, S. Pandya, C. Tucker, and C. Liu. Development and characterization of high sensitivity bioinspired artificial haircell sensor. In *Proceedings of Solid-State Sensors, Actuators, and Microsystems Workshop*, volume 6, pages 4–8. Citeseer, 2006.

[68] M. Asadnia, A. G. P. Kottapalli, K. D. Karavitaki, M. E. Warkiani, J. Miao, D. P. Corey, and M. Triantafyllou. From biological cilia to artificial flow sensors: Biomimetic soft polymer nanosensors with high sensing performance. *Scientific Reports*, 6:32955, 2016.

[69] A. T. Abdulsadda and X. Tan. Underwater source localization using an ipmc-based artificial lateral line. In *2011 IEEE International Conference on Robotics and Automation*, pages 2719–2724. IEEE, 2011.

[70] M. Asadnia, A. G. P. Kottapalli, J. Miao, M. E. Warkiani, and M. S. Triantafyllou. Artificial fish skin of self-powered micro-electromechanical systems hair cells for sensing hydrodynamic flow phenomena. *Journal of the Royal Society Interface*, 12(111):20150322, 2015.

[71] G. Krijnen, T. Lammerink, R. Wiegerink, and J. Casas. Cricket inspired flow-sensor arrays. In *SENSORS, 2007 IEEE*, pages 539–546. IEEE, 2007.

[72] J. B. Stocking, W. C. Eberhardt, Y. A. Shakhsheer, B. H. Calhoun, J. R. Paulus, and M. Appleby. A capacitance-based whisker-like artificial sensor for fluid motion sensing. In *SENSORS, 2010 IEEE*, pages 2224–2229. IEEE, 2010.

[73] J. J. Van Baar, M. Dijkstra, R. J. Wiegerink, T. S. J. Lammerink, and G. J. M. Krijnen. Fabrication of arrays of artificial hairs for complex flow pattern recognition. In *SENSORS, 2003 IEEE*, volume 1, pages 332–336. IEEE, 2003.

[74] N. Izadi, Meint. J. de Boer, J. W. Berenschot, and G. J. M. Krijnen. Fabrication of superficial neuromast inspired capacitive flow sensors. *Journal of Micromechanics and Microengineering*, 20(8):085041, 2010.

[75] A. Klein and H. Bleckmann. Determination of object position, vortex shedding frequency and flow velocity using artificial lateral

line canals. *Beilstein Journal of Nanotechnology*, 2(1):276–283, 2011.

[76] S. Große and W. Schröder. The micro-pillar shear-stress sensor mps3 for turbulent flow. *Sensors*, 9(4):2222–2251, 2009.

[77] B. J. Wolf, J. A. S. Morton, W. N. MacPherson, and S. M. van Netten. Bio-inspired all-optical artificial neuromast for 2d flow sensing. *Bioinspiration & Biomimetics*, 13(2):026013, 2018.

[78] S. Pandya, Y. Yang, D. L. Jones, J. Engel, and C. Liu. Multisensor processing algorithms for underwater dipole localization and tracking using mems artificial lateral-line sensors. *EURASIP Journal on Advances in Signal Processing*, 2006(1):076593, 2006.

[79] P. Liu, R. Zhu, and R. Que. A flexible flow sensor system and its characteristics for fluid mechanics measurements. *Sensors*, 9(12):9533–9543, 2009.

[80] Y. Yang, J. Chen, J. Engel, S. Pandya, N. Chen, C. Tucker, S. Coombs, D. L. Jones, and C. Liu. Distant touch hydrodynamic imaging with an artificial lateral line. *Proceedings of the National Academy of Sciences*, 103(50):18891–18895, 2006.

[81] J. Chen, J. Engel, N. Chen, S. Pandya, S. Coombs, and C. Liu. Artificial lateral line and hydrodynamic object tracking. In *19th IEEE International Conference on Micro Electro Mechanical Systems*, pages 694–697. IEEE, 2006.

[82] S. Verma, C. Papadimitriou, N. Lüthen, G. Arampatzis, and P. Koumoutsakos. Optimal sensor placement for artificial swimmers. *Journal of Fluid Mechanics*, 884, 2020.

[83] D. Xu, Z. Lv, H. Zeng, H. Bessaih, and B. Sun. Sensor placement optimization in the artificial lateral line using optimal weight analysis combining feature distance and variance evaluation. *ISA Transactions*, 86:110–121, 2019.

[84] R. Venturelli, O. Akanyeti, F. Visentin, J. Ježov, L. D. Chambers, G. Toming, J. Brown, M. Kruusmaa, W. M. Megill, and P. Fiorini. Hydrodynamic pressure sensing with an artificial lateral line in steady and unsteady flows. *Bioinspiration & Biomimetics*, 7(3):036004, 2012.

[85] T. Salumäe and M. Kruusmaa. Flow-relative control of an underwater robot. *Proceedings of the Royal Society A: Mathematical, Physical and Engineering Sciences*, 469(2153):20120671, 2013.

[86] J. F. Fuentes-Pérez, J. A. Tuhtan, R. Carbonell-Baeza, M. Musall, G. Toming, N. Muhammad, and M. Kruusmaa. Current velocity estimation using a lateral line probe. *Ecological Engineering*, 85:296–300, 2015.

[87] J. A. Tuhtan, J. F. Fuentes-Perez, G. Toming, M. Schneider, R. Schwarzenberger, M. Schletterer, and M. Kruusmaa. Man-made flows from a fish's perspective: autonomous classification of turbulent fishway flows with field data collected using an artificial lateral line. *Bioinspiration & Biomimetics*, 13(4):046006, 2018.

[88] G. Liu, S. Liu, S. Wang, H. Hao, and M. Wang. Research on artificial lateral line perception of flow field based on pressure difference matrix. *Journal of Bionic Engineering*, 16(6):1007–1018, 2019.

[89] N. Strokina, J.-K. Kämäräinen, J. A. Tuhtan, J. F. Fuentes-Pérez, and M. Kruusmaa. Joint estimation of bulk flow velocity and angle using a lateral line probe. *IEEE Transactions on Instrumentation and Measurement*, 65(3):601–613, 2015.

[90] J. A. Tuhtan, J. F. Fuentes-Perez, G. Toming, and M. Kruusmaa. Flow velocity estimation using a fish-shaped lateral line probe with product-moment correlation features and a neural network. *Flow Measurement and Instrumentation*, 54:1–8, 2017.

[91] G. Liu, H. Hao, T. Yang, S. Liu, M. Wang, A. Incecik, and Z. Li. Flow field perception of a moving carrier based on an artificial lateral line system. *Sensors*, 20(5):1512, 2020.

[92] Z. Ren and K. Mohseni. A model of the lateral line of fish for vortex sensing. *Bioinspiration & Biomimetics*, 7(3):036016, 2012.

[93] B. Free, M. K. Patnaik, and D. A. Paley. Observability-based path-planning and flow-relative control of a bioinspired sensor array in a karman vortex street. In *2017 American Control Conference (ACC)*, pages 548–554. IEEE, 2017.

[94] B. A. Free and D. A. Paley. Model-based observer and feedback control design for a rigid joukowski foil in a kármán vortex street. *Bioinspiration & Biomimetics*, 13(3):035001, 2018.

[95] H. Bleckmann. Reception of hydrodynamic stimuli in aquatic and semiaquatic animals. 1994.

[96] Z. Tang, Z. Wang, J. Lu, G. Ma, and P. Zhang. Underwater robot detection system based on fish's lateral line. *Electronics*, 8(5):566, 2019.

[97] X. Lin, Y. Zhang, M. Ji, X. Zheng, L. V. Kehong, J. Qiu, and G. Liu. Dipole source localization based on least square method and 3d printing. In *2018 IEEE International Conference on Mechatronics and Automation (ICMA)*, pages 2203–2208. IEEE, 2018.

[98] A. T. Abdulsadda and X. Tan. Nonlinear estimation-based dipole source localization for artificial lateral line systems. *Bioinspiration & Biomimetics*, 8(2):026005, 2013.

[99] A. Ahrari, H. Lei, M. A. Sharif, K. Deb, and X. Tan. Design optimization of an artificial lateral line system incorporating flow and sensor uncertainties. *Engineering Optimization*, 49(2):328–344, 2017.

[100] G. Liu, S. Gao, Th Sarkodie-Gyan, and Z. Li. A novel biomimetic sensor system for vibration source perception of autonomous underwater vehicles based on artificial lateral lines. *Measurement Science and Technology*, 29(12):125102, 2018.

[101] X. Chen, G. Zhu, X. Yang, D. L. S. Hung, and X. Tan. Model-based estimation of flow characteristics using an ionic polymer–metal composite beam. *IEEE/ASME Transactions on Mechatronics*, 18(3):932–943, 2012.

[102] A. T. Abdulsadda and X. Tan. Localization of a moving dipole source underwater using an artificial lateral line. In *Bioinspiration, Biomimetics, and Bioreplication 2012*, volume 8339, page 833909. International Society for Optics and Photonics, 2012.

[103] A. M. K. Dagamseh, T. S. J. Lammerink, C. M. Bruinink, R. J. Wiegerink, and G. J. M. Krijnen. Dipole source localisation using bio-mimetic flow-sensor arrays. *Procedia chemistry*, 1(1):891–894, 2009.

[104] A. M. K. Dagamseh, T. S. J. Lammerink, R. J. Wiegerink, and G. J. M. Krijnen. A simulation study of the dipole source localisation applied on bio-mimetic flow-sensor linear array. In *12th Annual Workshop on Semiconductor Advances for Future Electronics and Sensors (SAFE)*, pages 534–537. Technology Foundation (STW), 2009.

[105] A. M. K. Dagamseh and G. J. M. Krijnen. Map estimation of airflow dipole source positions using array signal processing. In *Annual Workshop on Semiconductor Advances for Future Electronics and Sensors, SAFE 2010*. Technology Foundation (STW), 2010.

[106] A. M. K. Dagamseh, T. S. J. Lammerink, M. L. Kolster, C. M. Bruinink, R. J. Wiegerink, and G. J. M. Krijnen. Dipole-source localization using biomimetic flow-sensor arrays positioned as lateral-line system. *Sensors and actuators A: Physical*, 162(2):355–360, 2010.

[107] A. Dagamseh, R. Wiegerink, T. Lammerink, and G. Krijnen. Imaging dipole flow sources using an artificial lateral-line system made of biomimetic hair flow sensors. *Journal of the Royal Society Interface*, 10(83):20130162, 2013.

[108] M. Ji, Y. Zhang, X. Zheng, X. Lin, G. Liu, and J. Qiu. Resolution improvement of dipole source localization for artificial lateral lines based on multiple signal classification. *Bioinspiration & Biomimetics*, 14(1):016016, 2018.

[109] M. Ji, Y. Zhang, X. Zheng, X. Lin, G. Liu, and J. Qiu. Performance evaluation and analysis for dipole source localization with lateral line sensor arrays. *Measurement Science and Technology*, 30(11):115107, 2019.

[110] B. J. Wolf and S. M. van Netten. Hydrodynamic imaging using an all-optical 2d artificial lateral line. In *2019 IEEE Sensors Applications Symposium (SAS)*, pages 1–6. IEEE, 2019.

[111] B. J. Wolf, S. Warmelink, and S. M. van Netten. Recurrent neural networks for hydrodynamic imaging using a 2d-sensitive artificial lateral line. *Bioinspiration & Biomimetics*, 14(5):055001, 2019.

[112] B. Wolf, P. Pirih, M. Kruusmaa, and S. M. van Netten. Shape classification using hydrodynamic detection via a sparse large-scale 2d-sensitive artificial lateral line. 2020.

[113] C. Wang, W. Wang, and G. Xie. Speed estimation for robotic fish using onboard artificial lateral line and inertial measurement unit. In *2015 IEEE International Conference on Robotics and Biomimetics (ROBIO)*, pages 285–290. IEEE, 2015.

[114] F. Zhang, F. D. Lagor, D. Yeo, P. Washington, and D. A. Paley. Distributed flow sensing for closed-loop speed control of a flexible fish robot. *Bioinspiration & Biomimetics*, 10(6):065001, 2015.

[115] N. Martiny, S. Sosnowski, K. Kühnlenz, S. Hirche, Y. Nie, J.-M. P. Franosch, and J. L. V. Hemmen. Design of a lateral-line sensor for an autonomous underwater vehicle. *IFAC Proceedings Volumes*, 42(18):292–297, 2009.

[116] O. Akanyeti, L. D. Chambers, J. Ježov, J. Brown, R. Venturelli, M. Kruusmaa, W. M. Megill, and P. Fiorini. Self-motion effects on hydrodynamic pressure sensing: part i. forward–backward motion. *Bioinspiration & Biomimetics*, 8(2):026001, 2013.

[117] M. Kruusmaa, P. Fiorini, W. Megill, M. de Vittorio, O. Akanyeti, F. Visentin, L. Chambers, H. El Daou, M.-C. Fiazza, J. Ježov, et al. Filose for svenning: A flow sensing bioinspired robot. *IEEE Robotics & Automation Magazine*, 21(3):51–62, 2014.

[118] X. Zheng, W. Wang, M. Xiong, and G. Xie. Online state estimation of a fin-actuated underwater robot using artificial lateral line system. *IEEE Transactions on Robotics*, 36(2):472–487, 2020.

[119] H. Liu, K. Zhong, Y. Fu, G. Xie, and Q. Zhu. Pattern recognition for robotic fish swimming gaits based on artificial lateral line system and subtractive clustering algorithms. *Sensors & Transducers*, 182(11):207, 2014.

[120] M. Kruusmaa, G. Toming, T. Salumäe, J. Ježov, and A. Ernits. Swimming speed control and on-board flow sensing of an artificial trout. In *2011 IEEE International Conference on Robotics and Automation*, pages 1791–1796. IEEE, 2011.

[121] T. Salumäe, I. Ranó, O. Akanyeti, and M. Kruusmaa. Against the flow: A braitenberg controller for a fish robot. In *2012 IEEE International Conference on Robotics and Automation*, pages 4210–4215. IEEE, 2012.

[122] W. Wang, Y. Li, X. Zhang, C. Wang, S. Chen, and G. Xie. Speed evaluation of a freely swimming robotic fish with an artificial lateral line. In *2016 IEEE International Conference on Robotics and Automation (ICRA)*, pages 4737–4742. IEEE, 2016.

[123] F. D. Lagor, L. D. DeVries, K. M. Waychoff, and D. A. Paley. Bio-inspired flow sensing and control: Autonomous underwater navigation using distributed pressure measurements. In *Proc. 18th Int. Symp. Unmanned Untethered Submersible Technol.*, pages 1–8, 2013.

[124] W. Wang, X. Zhang, J. Zhao, and G. Xie. Sensing the neighboring robot by the artificial lateral line of a bio-inspired robotic fish. In *2015 IEEE/RSJ International Conference on Intelligent Robots and Systems (IROS)*, pages 1565–1570. IEEE, 2015.

[125] N. Muhammad, N. Strokina, G. Toming, J. Tuhtan, J.-K. Kämäräinen, and M. Kruusmaa. Flow feature extraction for underwater robot localization: Preliminary results. In *2015 IEEE International Conference on Robotics and Automation (ICRA)*, pages 1125–1130. IEEE, 2015.

[126] J. F. Fuentes-Pérez, N. Muhammad, J. A. Tuhtan, R. Carbonell-Baeza, M. Musall, G. Toming, and M. Kruusmaa. Map-based localization in structured underwater environment using simulated hydrodynamic maps and an artificial lateral line. In *2017 IEEE International Conference on Robotics and Biomimetics (ROBIO)*, pages 128–134. IEEE, 2017.

[127] M. Bariche. First record of the cube boxfish ostracion cubicus (ostraciidae) and additional records of champsodon vorax (champsodontidae) from the mediterranean. *Aqua*, 17(17):181–184, 2011.

[128] W. Wang and G. Xie. Cpg-based locomotion controller design for a boxfish-like robot. *International Journal of Advanced Robotic Systems*, 11(87):147–169, 2014.

[129] X. Zheng, W. Wang, C. Wang, R. Fan, and G. Xie. An introduction of vision-based autonomous robotic fish competition. In *The 12th World Congress on Intelligent Control and Automation*, pages 2561–2566, Guilin, China, June 2016.

[130] M. Nakae and K. Sasaki. The lateral line system and its innervation in the boxfish ostracion immaculatus (tetraodontiformes: Ostraciidae): description and comparisons with other tetraodontiform and perciform conditions. *Ichthyological Research*, 52(4):343–353, 2005.

[131] J. J. Leonard and H. F. Durrant-Whyte. Mobile robot localization by tracking geometric beacons. *IEEE Transactions on robotics and Automation*, 7(3):376–382, July 1991.

[132] X. Yun, E. R. Bachmann, R. B. Mcghee, R. H. Whalen, R. L. Roberts, R. G. Knapp, A. J. Healey, and M. J. Zyda. Testing and evaluation of an integrated gps/ins system for small auv navigation. *IEEE Journal of Oceanic Engineering*, 24(3):396–404, 1999.

[133] M. J. Stanway. Water profile navigation with an acoustic doppler current profiler. In *OCEANS'10 IEEE SYDNEY*, pages 1–5. IEEE, 2010.

[134] A. Elfes. Sonar-based real-world mapping and navigation. *IEEE Journal on Robotics and Automation*, 3(3):249–265, 1987.

[135] W. Burgard, A. Derr, D. Fox, and A. B. Cremers. Integrating global position estimation and position tracking for mobile robots: the dynamic markov localization approach. In *IEEE/RSJ International Conference on Intelligent Robots and Systems*, Victoria, BC, Canada, October 1998.

[136] J. Lighthill. Estimates of pressure differences across the head of a swimming clupeid fish. *Philosophical Transactions of the Royal Society of London. Series B: Biological Sciences*, 341(1296):129–140, 1993.

[137] G. Zhou, Y. Zhao, and F. Guo. A temperature compensation system for silicon pressure sensor based on neural networks. In *Nano/Micro Engineered and Molecular Systems (NEMS), 2014 9th IEEE International Conference on*, pages 467–470, Waikiki Beach, HI, USA, April 2014. IEEE.

[138] W. Xie and P. Bai. A pressure sensor calibration model based on support vector machine. In *Control and Decision Conference (CCDC), 2012 24th Chinese*, pages 3239–3242, Taiyuan, China, May 2012. IEEE.

[139] A. T. Abdulsadda. *Artificial lateral line systems for feedback control of underwater robots.* Michigan State University, 2012.

[140] A. T. Abdulsadda and X. Tan. Localization of source with unknown amplitude using ipmc sensor arrays. *SPIE*, 7976(10):2378–2384, 2011.

[141] A. T. Abdulsadda and X. Tan. Underwater tracking of a moving dipole source using an artificial lateral line: algorithm and experimental validation with ionic polymer–metal composite flow sensors. *Smart Materials and Structures*, 22(4):045010, 2013.

[142] W. K. Yen, S. D. Martinez, and J. Guo. Controller design for a fish robot to follow an oscillating source. In *IEEE International Conference on Cyber Technology in Automation, Control, and Intelligent Systems*, pages 959–964, Shenyang, China, June 2015.

[143] J. C. Nauen and G. V. Lauder. Hydrodynamics of caudal fin locomotion by chub mackerel, scomber japonicus (scombridae). *Journal of Experimental Biology*, 205(12):1709–1724, 2002.

[144] J.-M. P. Franosch, H. J. A. Hagedorn, J. Goulet, J. Engelmann, and J. L. van Hemmen. Wake tracking and the detection of vortex rings by the canal lateral line of fish. *Physical Review Letters*, 103(7):078102, 2009.

[145] G. V. Lauder, E. J. Anderson, J. Tangorra, and P. G. Madden. Fish biorobotics: kinematics and hydrodynamics of self-propulsion. *Journal of Experimental Biology*, 210(16):2767–2780, 2007.

[146] X. Zhang, W. Wang, S. Chen, and G. Xie. Study on perception of neighbor bionic fish based on artificial lateral line system (in chinese). *Measurement & Control Technology*, 35(10):33–37, 2016.

[147] M. A. Green, C. W. Rowley, and A. J. Smits. The unsteady three-dimensional wake produced by a trapezoidal pitching panel. *Journal of Fluid Mechanics*, 685(7):117–145, 2011.

[148] D. Bernoulli. *Hydrodynamica: sive de viribus et motibus fluidorum commentarii.* Johann Heinrich Decker, Strasbourg, 1738.

[149] I. K. Bartol, M. Gharib, P. W. Webb, D. Weihs, and M. S. Gordon. Body-induced vortical flows: a common mechanism for self-corrective trimming control in boxfishes. *Journal of Experimental Biology*, 208(2):327–344, 2005.

[150] I. K. Bartol, M. S. Gordon, P. Webb, D. Weihs, and M. Gharib. Evidence of self-correcting spiral flows in swimming boxfishes. *Bioinspiration & Biomimetics*, 3(1):014001, 2008.

[151] P. Kodati and X. Deng. Towards the body shape design of a hydrodynamically stable robotic boxfish. In *IEEE/RSJ International Conference on Robotics and Automation*, pages 5412–5417, Beijing, China, October 2006.

[152] O. Akanyeti, P. J. Thornycroft, G. V. Lauder, Y. R. Yanagitsuru, A. N. Peterson, and J. C. Liao. Fish optimize sensing and respiration during undulatory swimming. *Nature Communications*, 7:11044, 2016.

[153] W. J. Stewart, F. B. Tian, O. Akanyeti, C. J. Walker, and J. C. Liao. Refuging rainbow trout selectively exploit flows behind tandem cylinders. *Journal of Experimental Biology*, 219(14):2182–2191, 2016.

[154] H. W. Lissman and K. E. Machin. The mechanism of object location in gymnarchus niloticus and similar fish. *Journal of Experimental Biology*, 35(2):451–486, 1958.

[155] P. Moller. *Electric fishes: history and behavior.* Chapman & Hall, London, 1995.

[156] J. R. Solberg, K. M. Lynch, and M. A. Maciver. Active electrolocation for underwater target localization. *The International Journal of Robotics Research*, 27(5):529–548, 2008.

[157] B. Yang, J. B. Snyder, M. Peshkin, and M. A. Maciver. Finding and identifying simple objects underwater with active electrosense. *The International Journal of Robotics Research*, 34(10):1255–1277, 2015.

[158] C. Chevallereau, M. R. Benachenhou, V. Lebastard, and F. Boyer. Electric sensor-based control of underwater robot groups. *IEEE Transactions on Robotics*, 30(3):604–618, 2014.

[159] D. Weihs. Some hydrodynamical aspects of fish schooling. In *Swimming and flying in nature*, pages 703–718. Springer, 1975.

[160] X. Zheng, C. Wang, R. Fan, and G. Xie. Artificial lateral line based local sensing between two adjacent robotic fish. *Bioinspiration & Biomimetics*, 13(1):016002, November 2018.

[161] O. Akanyeti, R. Venturelli, F. Visentin, L. Chambers, W. M. Megill, and P. Fiorini. What information do kármán streets offer to flow sensing? *Bioinspiration & Biomimetics*, 6(3):036001, 2011.

[162] L. DeVries and D. A. Paley. Observability-based optimization for flow sensing and control of an underwater vehicle in a uniform flowfield. In *2013 American Control Conference*, pages 1386–1391. IEEE, 2013.

[163] L. Breiman. Random forests. *Machine Learning*, 45(1):5–32, 2001.

[164] Y. Liu, Y. Wang, and J. Zhang. New machine learning algorithm: Random forest. In *International Conference on Information Computing and Applications*, pages 246–252. Springer, 2012.

[165] K.-l. Hsu, H. V. Gupta, and S. Sorooshian. Artificial neural network modeling of the rainfall-runoff process. *Water Resources Research*, 31(10):2517–2530, 1995.

[166] R. Hecht-Nielsen. Theory of the backpropagation neural network. In *Neural networks for perception*, pages 65–93. Elsevier, 1992.

[167] S. R. Gunn et al. Support vector machines for classification and regression. *ISIS Technical Report*, 14(1):5–16, 1998.

[168] J. D. Anderson Jr. *Fundamentals of aerodynamics*. Tata McGraw-Hill Education, 2010.

Index

Note: Locators in *italics* represent figures and **bold** indicate tables in the text.

Milton Keynes UK
Ingram Content Group UK Ltd.
UKHW031132141024
449569UK00006B/234

9 781032 316161